Enteric Viruses in Aquatic Environments

Enteric Viruses in Aquatic Environments

Special Issue Editors

Eiji Haramoto
Masaaki Kitajima

MDPI • Basel • Beijing • Wuhan • Barcelona • Belgrade • Manchester • Tokyo • Cluj • Tianjin

Special Issue Editors
Eiji Haramoto
University of Yamanashi
Japan

Masaaki Kitajima
Hokkaido University
Japan

Editorial Office
MDPI
St. Alban-Anlage 66
4052 Basel, Switzerland

This is a reprint of articles from the Special Issue published online in the open access journal *Pathogens* (ISSN 2076-0817) (available at: https://www.mdpi.com/journal/pathogens/special_issues/Enteric_Viruses_Aquatic_Environments).

For citation purposes, cite each article independently as indicated on the article page online and as indicated below:

LastName, A.A.; LastName, B.B.; LastName, C.C. Article Title. *Journal Name* **Year**, *Article Number*, Page Range.

ISBN 978-3-03928-568-6 (Pbk)
ISBN 978-3-03928-569-3 (PDF)

Contents

About the Special Issue Editors vii

Preface to "Enteric Viruses in Aquatic Environments" ix

Charles P. Gerba and Walter Q. Betancourt
Assessing the Occurrence of Waterborne Viruses in Reuse Systems: Analytical Limits and Needs
Reprinted from: *Pathogens* **2019**, *8*, 107, doi:10.3390/pathogens8030107 1

Bikash Malla, Rajani Ghaju Shrestha, Sarmila Tandukar, Dinesh Bhandari, Ocean Thakali, Jeevan B. Sherchand and Eiji Haramoto
Detection of Pathogenic Viruses, Pathogen Indicators, and Fecal-Source Markers within Tanker Water and Their Sources in the Kathmandu Valley, Nepal
Reprinted from: *Pathogens* **2019**, *8*, 81, doi:10.3390/pathogens8020081 13

Erika Ito, Jian Pu, Takayuki Miura, Shinobu Kazama, Masateru Nishiyama, Hiroaki Ito, Yoshimitsu Konta, Gia Thanh Nguyen, Tatsuo Omura and Toru Watanabe
Weekly Variation of Rotavirus A Concentrations in Sewage and Oysters in Japan, 2014–2016
Reprinted from: *Pathogens* **2019**, *8*, 89, doi:10.3390/pathogens8030089 27

Koichi Matsubara and Hiroyuki Katayama
Development of a Portable Detection Method for Enteric Viruses from Ambient Air and Its Application to a Wastewater Treatment Plant
Reprinted from: *Pathogens* **2019**, *8*, 131, doi:10.3390/pathogens8030131 33

Zheng Ji, Xiaochang C. Wang, Limei Xu, Chongmiao Zhang, Cheng Rong, Andri Taruna Rachmadi, Mohan Amarasiri, Satoshi Okabe, Naoyuki Funamizu and Daisuke Sano
Fecal Source Tracking in A Wastewater Treatment and Reclamation System Using Multiple Waterborne Gastroenteritis Viruses
Reprinted from: *Pathogens* **2019**, *8*, 170, doi:10.3390/pathogens8040170 45

Suntae Lee, Mamoru Suwa and Hiroyuki Shigemura
Metagenomic Analysis of Infectious F-Specific RNA Bacteriophage Strains in Wastewater Treatment and Disinfection Processes
Reprinted from: *Pathogens* **2019**, *8*, 217, doi:10.3390/pathogens8040217 59

About the Special Issue Editors

Eiji Haramoto is a Professor at Interdisciplinary Center for River Basin Environment, University of Yamanashi, Japan. Dr. Haramoto received his Doctor of Engineering degree at Department of Urban Engineering, the University of Tokyo, Japan, in March 2007. Between April 2007 and August 2008, he worked at the Department of Water Supply Engineering, National Institute of Public Health, Japan, as a research fellow of Japan Society for the Promotion of Science, and then he was employed at University of Yamanashi as an Assistant Professor in September 2008 and promoted to an Associate Professor in January 2015 and to a Professor in March 2020. His research interests include the development of concentration/detection methods for waterborne pathogens (viruses, protozoa, and bacteria), the spatial and temporal prevalence of waterborne pathogens and their indicators in aquatic environments, the reduction of waterborne pathogens during water/wastewater treatment processes, and microbial source tracking using host-specific microbial genetic markers. Dr. Haramoto has published more than 100 peer-reviewed papers and is now an Editorial Board Member of Applied and Environmental Microbiology (AEM) and Water (MDPI).

Masaaki Kitajima earned his Doctor of Engineering degree in the field of urban environmental engineering from the University of Tokyo in 2011. After receiving the doctoral degree, he worked at the University of Arizona as a Post-Doc through the Japan Society for Promotion of Science Postdoctoral Fellowships for Research Abroad from 2011 to 2013. He expanded his research experience by working at the Singapore-MIT Alliance for Research and Technology Center, a research center of MIT in Singapore, for 2 years from 2014. Since 2016, he is an Assistant Professor at Hokkaido University in Japan, where he teaches courses related to environmental engineering and mentors undergraduate and graduate students. Dr. Kitajima has published 81 peer-reviewed journal papers in his career. His research area covers broad aspects of health-related water microbiology from an environmental engineering perspective, represented by tracking the behavior of microbial pathogens, and indicators in natural and engineered water systems.

Preface to "Enteric Viruses in Aquatic Environments"

Enteric viruses, such as noroviruses, adenoviruses, and rotaviruses, are excreted into feces of the infected individuals and can be transmitted through a fecal-oral route via contaminated food and water. Thus, it is important to understand the prevalence of enteric viruses in aquatic environments, along with their behaviors during water and wastewater treatment processes. The development of methods for concentrating, detecting, and quantifying enteric viruses in environmental samples is still a challenging issue as no gold standard methods have been established. Recent viral metagenomic studies have demonstrated great genetic diversity of enteric viruses, identifying novel viruses. Studies on indicators of enteric viruses and even on viral indicators of fecal contamination are also necessary for better management of microbial water quality.

This Special Issue on "Enteric viruses in aquatic environments" addresses cutting-edge research and review articles from leading scientists in the field of water and environmental virology.

Eiji Haramoto, Masaaki Kitajima
Special Issue Editors

pathogens

MDPI

Review

Assessing the Occurrence of Waterborne Viruses in Reuse Systems: Analytical Limits and Needs

Charles P. Gerba and Walter Q. Betancourt *

Water and Energy Sustainable Technology (WEST) Center, The University of Arizona, 2959 W. Calle Agua Nueva, Tucson, AZ 85745, USA

* Correspondence: wbetancourt@email.arizona.edu; Tel.: +1-(520)-621-6163

Received: 16 June 2019; Accepted: 19 July 2019; Published: 22 July 2019

Abstract: Detection of waterborne enteric viruses is an essential tool in assessing the risk of waterborne transmission. Cell culture is considered a gold standard for detection of these viruses. However, it is important to recognize the uncertainty and limitations of enteric virus detection in cell culture. Cell culture cannot support replication of all virus types and strains, and numerous factors control the efficacy of specific virus detection assays, including chemical additives, cell culture passage number, and sequential passage of a sample in cell culture. These factors can result in a 2- to 100-fold underestimation of virus infectivity. Molecular methods reduce the time for detection of viruses and are useful for detection of those that do not produce cytopathogenic effects. The usefulness of polymerase chain reaction (PCR) to access virus infectivity has been demonstrated for only a limited number of enteric viruses and is limited by an understanding of the mechanism of virus inactivation. All of these issues are important to consider when assessing waterborne infectious viruses and expected goals on virus reductions needed for recycled water. The use of safety factors to account for this may be useful to ensure that the risks in drinking water and recycled water for potable reuse are minimized.

Keywords: virus; infectivity; cell culture; molecular methods; wastewater; reuse

1. Introduction

Quantifying the number of infectious viruses in water and wastewater is necessary to determine the risks associated with exposure (e.g., ingestion) and in determining the degree of treatment needed to reduce these risks to an acceptable level [1–3]. For example, the state of California requires a 12-$\log_{10}$ reduction of all human enteric viruses in recycled waters for potable reuse applications [4]. To achieve this goal, knowledge of the number of infectious viruses in wastewater before treatment is needed. Infectious viruses are defined as those capable of replicating in cell culture and thus, have the potential to replicate in humans and animals and cause disease. In this review, the term infectivity is used in reference to the ability of methods to measure infectious viruses. This requires methods that can determine the number of infectious viruses. The purpose of this review is to provide an understanding of the limitations of current methods for assessing the infectivity of waterborne enteric viruses. We believe that this is essential for interpreting the data on viruses in water for persons involved in assessing needed technology for the treatment of recycled water for reuse applications while considering the associated risks.

Before the development and application of molecular methods for the assessment of virus occurrence in water, animal cell culture was the only practical method available. Virus growth in cell culture indicates the potential for the virus to replicate in humans and cause disease. Enteroviruses were found to readily grow in cell culture from the earliest days of techniques for maintaining animal cells in the laboratory. Because they were so easily cultivated, most of our historic knowledge on enteric virus behavior in water and removal by water/wastewater treatment processes is based on enteroviruses. The safety associated with vaccine poliovirus strains allowed for bench- and pilot-scale testing of treatment processes under controlled conditions. However, enteroviruses have rarely been associated with waterborne disease, and today, we know they are only a small fraction of the viral community found in wastewater that is capable of causing illness in humans [3,5]. This has been in part revealed through the application of the quantitative polymerase chain reaction (qPCR) assay and more recently by viral sewage metagenomics [6–9]. Unfortunately, these methods cannot directly detect the infectivity of waterborne viruses. Various approaches have been developed to assess infectivity of waterborne enteric viruses using molecular methods, but they are specific to the virus and the mechanism of virus inactivation [10–12]. The mechanism of virus inactivation may vary by the type of virus, disinfectant, and other methods which may make the virus incapable of replication [13,14]. Thus, there is no universal method which can substitute for cell culture assessing viral infectivity in humans and animals.

2. Factors Affecting Virus Infectivity in Cell Culture

2.1. Type of Cell Culture (Continuous vs. Primary)

Two types of cell culture have been used for the detection of viruses in water. Primary cell cultures originate directly from the organs of animals and humans. The most commonly used cell cultures in virology derive from primates, including humans and monkeys; rodents, such as hamsters, rats, and mice; and birds, most notably chickens [15,16]. Moreover, cells from a primary culture may be subcultured to obtain a large number of cells. Cultures established in this fashion from primary cell cultures are called secondary cultures [16]. They can only be passaged for a limited number of cell generations (usually 20 to 100) after which the cells cease to divide, then degenerate and die, a phenomenon called crisis or senescence. On the other hand, continuous cell lines may be passaged indefinitely as they originate from transformed cells that are no longer subject to senescence. Continuous cell lines are relatively easy to maintain because they can be passaged indefinitely and are the cell line of choice today for environmental virology research. Primary cells from human and nonhuman primates are the most sensitive to the widest variety of viruses which infect humans since these cells maintain many of the important markers and functions seen in vivo [17,18]. However, primary cells are not in common use today. Continuous cell lines from human and nonhuman primates are usually more restrictive to the types of viruses they can propagate (Table 1). This is because the cell surface must have specific receptors for the attachment and replication of the virus. The continuous cell line, Buffalo green monkey (BGM), was selected for use by the United States Environmental Protection Agency because certain coxsackieviruses (CV) and polioviruses (PV) grew well in this cell line, producing cytopathogenic effects with similar sensitivity to virus growth in primary cells [19]. BGM cells were found to be the most sensitive continuous cell line for the detection of enteroviruses [20] and have become the most commonly used cell line for the detection of enteric viruses in water and wastewater in the United States for over 30 years. While laboratory strains of echovirus will grow in this cell line, its use with environmental samples tends to favor the isolation of group B coxsackieviruses [20,21]. This may be due to the more rapid growth of group B coxsackieviruses in BGM cells [20]. An exhaustive comparison of cell lines and enteric virus susceptibility (16 cell lines against 105 different virus types) demonstrated a great deal of variability in cell susceptibility to virus type [22]. They found that not a single cell line could detect all enteroviruses, even of the same genus. In addition, with the control of poliovirus infections in the developed world and the elimination of the oral live poliovirus vaccine, vaccine strains of poliovirus are now absent in wastewater in most developed countries. Because live attenuated viruses replicate in the gut of vaccine recipients and spread person to person within a community, poliovirus was a common isolate in wastewater and sewage-polluted waters when vaccination was common from the mid-1950s until the mid-1990s in the

United States. This is not surprising as vaccine poliovirus strains were selected for their ability to grow in high titers in cell culture.

Table 1. Susceptibilities of cell culture lines most commonly used for isolation and detection of waterborne enteric viruses.

Cell Line	ADENO	CV-A	CV-B	ECHO	PV	REO	ROTAV	ASTROV
Human Embryonic Kidney	++	+	+		+	+		
A549	++++							
Buffalo Green Monkey (BGM)	+	+	++++	++	+++	+++		
Human rhabdomyosarcoma	-	++	-	++	++			
Caco-2 *	+	+	++	?	+	+	+	+
PLC/PRC/5 **	++		++	++?				
HEL-299 ***	++	++	++	-	+			
RD	+	++	-	++	+	+		

Note: The number of + signs indicate the relative degree of replication of the virus in the specific cell line. A "–"sign indicates no replication. *[23] **[24,25] ***[26] ADENO: Adenovirus; CV-A: Coxsackievirus A; CV-B: Coxsackievirus B; ECHO: Echovirus; PV: Poliovirus; REO: Reovirus, ROTAV: Rotavirus; ASTROV: Astrovirus, RD: Rhabdomyosarcoma titers in cell culture. Thus, it should be recognized that much of our information on viruses in water and the effectiveness of treatment processes comes from a very limited group of enteroviruses. The question mark indicates potential replication of the virus in the corresponding cell line.

The number of times a cell line has been passaged in the laboratory may also affect the ability of the virus to replicate (Table 2). Certain variants of the cells may be selected for over time because of their more rapid growth, which may be less or non-permissive to the replication of the virus.

Table 2. Factors that influence the infectivity of viruses in cell culture.

Factor	Remarks	References
Type of virus	Not all viruses can be grown in cell culture	[26]
Type of cell line	Not all viruses can be grown in the same cell culture	[26]
Number of times cell line has been passed in the laboratory	Cells may lose their sensitivity to virus infectivity after prolong passage in the laboratory; this may be virus-specific	[26,27]
Laboratory grown versus naturally occurring viruses	Laboratory grown viruses have been adapted for rapid growth and infectivity in cell culture.	[28]
Effectiveness of host cell repair enzymes	Host cell repair enzymes can repair damage to double-stranded DNA viruses after exposure to UV light. This may vary with cell line	[29]
Observation time for production of CPE	This may take days to weeks	[30]

Over time cell cultures become less efficient for replication of certain types of viruses [26]. Previous studies reported that BGM cells became less efficient to coxsackievirus B3 (CVB3) and CVB4 but were still sensitive to poliovirus 1 [27]. This questions the use of positive virus controls in environmental assays. Use of any specific strain of a laboratory grown virus does not mean the cell line has not lost its ability to replicate viruses of the same group or naturally occurring viruses of the same type.

In the case of the double-stranded DNA adenovirus, it has been found that replication of the virus after ultraviolet light exposure is dependent upon the ability of the cell line to repair damage to the DNA [31]. UV light causes crosslinking of the DNA and can be repaired by enzymes in the host cell. This ability depends on the cell line, with some being more effective than others [31].

2.2. *Cytotoxicity in Cell Lines*

Virus concentrates from different water matrices (e.g., surface water, sewage, secondary or tertiary treated wastewater, groundwater) can contain compounds toxic to cell cultures used for the detection of infectious viruses. The cytotoxicity may be associated with metals, complex mixtures of compounds associated with microalgae or plants as well as the reagents used for virus concentration and recovery [32–34]. Numerous methods have been applied for reducing cytotoxicity associated with the produced concentrates including sample dilution, washing cell monolayers with saline solution after inoculation, freon extraction, and cationic polyelectrolyte precipitation or high-speed centrifugation followed by filtration of the samples through positively charged depth filters [35–37]. Studies have also revealed that sample concentrates toxic to cells may not be necessarily inhibitory to the RT-PCR analysis [38].

2.3. *Virus Type*

Viral growth in cell culture is limited by the ability of virions to attach to specific receptors on the surface of animal cells and their ability to replicate within the cells. For the enteroviruses and many of the enteric viruses, this results in morphological changes induced in individual cells or groups of cells by virus infection that can be easily recognized by light microscopy and collectively called cytopathic or cytopathogenic effect (CPE). However, there are viruses whose replication may be limited to one or a few adjacent cells with no obvious cytopathogenic effects [15]. Alternative approaches, such as immunofluorescence, immunoperoxidase, electron microscopy or polymerase chain reaction (PCR) assays, have been used for detection of viruses that produce CPE slowly or not at all in cultured cells [15,28]. How this limited growth can be equated to the risk of infection and illness in humans is uncertain. Continuous cell lines are not necessarily reflective of the cells within the human host and the ability of the virus to destroy cells or establish themselves as subclinical or latent infections. For example, coxsackieviruses may establish lifelong latent infections in humans [39].

Cell culture is also less permissive for the growth of naturally occurring viruses than laboratory-grown viruses. Viruses grown in the laboratory have been selected for their rapid growth in cell culture and the number of virions observed under an electron microscope versus number observed by CPE or plaque-forming units (PFU) is usually 1:2 to 1:100 depending on the virus and method of assay [40,41]. In the case of naturally occurring viruses in stool samples, this ratio may be as great as 1:46,000 [42,43]. The particle-to-PFU ratio of poliovirus ranges from 30 to 1000, which is similar for other members of the *Picornaviridae* family [42]. Passage of naturally occurring viruses in cultured cells usually results in the significant lowering of this ratio [43] as mutants, which replicate in the specific cell line selected. Comparing the ratio of viral particles to genomes detected by molecular methods has also been attempted. However, several limitations exist, e.g., not all virus types grow in one type of cell culture, and there are differences in the quantitative precision of the methods for estimation of virus particles and viral genomes. Previous studies attempted to determine the ratio of enteroviruses detected by reverse transcription-PCR versus the number of infectious viruses determined in the cell culture [44]. The ratio of virus genomes to infectious virions reported in the study was 1:200. This ratio is likely significantly underestimated because the cells were only observed for 5 days for CPE, and not all enteroviruses can grow in this cell line. Another study comparing integrated cell culture-PCR (ICC-PCR) and real-time quantitative PCR (qPCR) in sewage polluted waters found that greater numbers of adenoviruses were detected by ICC-PCR [45]. Using the improved cell line (293 CMV) for detecting enteric human adenoviruses (HAdVs), the replication of HAdV in the cell line was determined by measuring the production of viral mRNA and determining the levels of viral DNA [46]. The results of the study demonstrated the effectiveness of the new transactivated 293 CMV cell line for improved propagation and detection of HAdVs from environmental samples. The ratio of infectious adenovirus with the improved cell line varied from 1:13.7 to 1:22 [46]. In a similar study, it was found that the ratio of infectious adenovirus by cell culture infectivity determined by the detection of viral mRNA production varied from 1:11 to 1:381 in untreated sewage [47].

The degree of viral aggregation may also influence the underestimation of infective viruses in a sample. Aggregated viruses in cell culture are often only counted as one infectious virus as a result of only one countable plaque [48]. However, they may represent thousands of potentially infectious viruses and have a greater probability of infection when ingested [49].

Another complicating factor is that one group of viruses may grow faster than another or interfere with the replication of another group of viruses [50], which again underestimates the true number of infectious viruses able to replicate in one specific cell line.

3. Impact of Assay Methods on Virus Detection

The three most common methods for quantitative detection of virus replication in cell culture are the total culturable virus quantal assay (TCVQA) which requires computation of a most probable number (MPN), the plaque assay which quantifies the number of plaque-forming units in a virus sample as plaque-forming units (PFU), and the 50% tissue culture infective dose ($TCID_{50}$) assay that quantifies the amount of virus required to produce CPE in, or kill 50% of virus-inoculated cultured cells in a multi-welled plate [16,51]. The TCVQA has been used for detection of enteric viruses in wastewater, but not all viruses will plaque or may require mixed cell types or pretreatment of cells before inoculation to form plaques [24]. Other limitations of the method include the difficulty to keep the monolayers beyond 5 to 7 days under an agar overlay, inability to perform a second passage, and laboratory strains which produce CPE in cell culture may not form plaques [21]. Numerous methods have been developed to determine the replication of viruses within cell culture (Table 3). None of these methods can detect all of the infective viruses in an environmental sample, even if the cell line is susceptible to the virus. As a first step, the virus must come into contact with a receptor on the cell membrane. Thus, the size of the inoculum (e.g., volume) of the sample, as well as a means to enhance contact with the cell membrane, are important steps in the efficiency of the assay for detecting infectious viruses. A previous study found that the optimal inoculum volume for poliovirus type 1 was one mL per 25 cm^2 of cell monolayer [28]. A marked decrease in the number of plaques was observed when over 1 mL of sample was inoculated on this surface area. The numbers of infectious viruses can also be increased by using roller bottles [52]. Secondary passage on fresh cells, use of suspended cell culture, rotating or shaking the liquid in the cell culture flasks during incubation [53], and use of suspended-cell may increase the number of viruses detected. All of these methods increase the probability of contact of the virus with receptors on the cell surface, i.e., the suspended virus must come into contact with cells. However, the increase in titer or probability of isolation may be virus and type dependent. For example, the suspended cell culture technique was found to increase the titer of poliovirus type 1 almost 10 fold but had no significant effect on echovirus 1 titer in BGM cells [15]. The appearance of CPE also varies greatly with naturally occurring viruses taking longer than laboratory-grown viruses. This is because laboratory-grown viruses have been selected for rapid growth in cell culture, as previously discussed. While CPE for vaccine strain of poliovirus may take only 48 hours, natural isolates of other enteroviruses may take five days or longer. A previous study demonstrated that going from a two-week incubation to three weeks resulted in a 100-fold increase titer in adenovirus 2 [30]. In the case of adenovirus 2 exposed to UV light, the increase in titer was 140-fold. This suggests that the longer incubation period allows for greater time for the cell enzymes to repair UV light damage of adenovirus. The most common methods to assess viral infectivity are shown in Table 3. All of these involve the use of cell culture except PCR. The use of the plaque-forming unit method previously mentioned, which involves an agar overlay of the cell monolayer to reduce virus spreading, results in a more precise quantification of viruses able to form plaques. This is true of naturally occurring viruses which can require a second passage or even a third passage before the production of CPE.

A variety of additives (Table 4) have been used to enhance viral infectivity in cell culture and to increase the range of susceptibility to a greater range of viral types [22]. For example, use of 5-ido-2′-deoxyuridine will result in plaque formation of adenovirus 1 and echoviruses [24].

Incorporation of enzymes is also known to enhance the infectivity of reoviruses in cell culture [40]. A secondary passage of environmental samples is often necessary for observation of CPE for some viruses [28]. This is because of the slower growth of naturally occurring viruses in cell lines and that not all the viruses in the sample will come into contact with the monolayer. Additional studies indicated that removal of the inoculum of poliovirus 1 from a cell flask containing a monolayer onto a fresh monolayer resulted in a 10-fold increase in titer of the virus (2300 to 24,000 most probable number) [41]. Passage a third time resulted in an additional increase in titer. For example, we have never observed the production of reovirus CPE in BGM cells in wastewater samples until a second or third passage [54]. Techniques using antibodies and PCR have the advantage of detecting virus replication without the production of CPE. Specific antibodies or primers are required for virus detection. PCR has the advantage in that replication of groups of viruses (i.e., enteroviruses, adenoviruses, rotavirus, etc.) can be identified, even if limited replication has taken place. Pool sera from multiple individuals has been used to determine the replication of viruses by an immunoperoxidase method, which increased the scope of viruses replicating in cell culture [28].

Table 3. Most common methods used to assess viral infectivity in cell culture from environmental samples.

Method	Reference
Plaque form unit (PFU)	[28]
Most probable number by cytopathogenic effects (MPN)	[28]
Tissue culture infectious dose 50% ($TCID_{50}$)	[28]
Integrated Cell Culture polymerase chain reaction (ICC-PCR)	[55]
Detection of messenger RNA	[47]

Table 4. Methods used to enhance cell culture infectivity to increase in virus numbers quantified.

Method	Virus	Increase in titer	Reference
5-iodo-2′-deoxyuridine	enteroviruses	Range of 0.7- to 3.3-fold increase in titer depending on virus type	[24]
Suspended cell culture agar method	Various enterovirus, wastewater and river water, and filtered water	Average 5.6-fold by plaque-forming unit method; range 0.1 to 23.3; 10 to 100 fold with BGM cells with polluted river water	[21,26]
Double agar overlay	Various enteroviruses and sewage isolates	7.7- to 12-fold over monolayer depending upon the virus	[56]
Rocking	Poliovirus 1	16% to 23% more plaques for rocked flasks. Increases rate of virus adsorption to cells	[53]
Adsorption time	Poliovirus 1	Maximum at 2 hours: ~2.5-fold from 30 min to 2 hrs for rocked flasks	[53]
Soluble proteins	Poliovirus 1	80% reduction of plaques in the presence of 3% beef extract compare to phosphate-buffered saline	[53]
Size of inoculum	Poliovirus 1	Inoculum of greater than one ml/25 cm^2 resulted in decreasing numbers of plaques and MPN	[28,53]
Flask vs. roller bottlePlaques vs. CPE	Poliovirus 1	Greater sensitivity when detecting low levels of virus in a sample	[52]
Sequential passage	Poliovirus 1	Titer can increase by 10- to 100-fold	[41,52]

4. Molecular Methods for Assessing Virus Infectivity

4.1. Integrated Cell Culture-Polymerase Chain Reaction

Various molecular methods have been developed to more rapidly determine the growth of viruses in cell culture and ultimately for the detection of viruses, which may not produce CPE or exhibit limited growth [55,57–59]. This still requires that each inoculated cell culture flask be tested and that primers for each group of virus to be tested are available. The major advantage of this method is that replicating viruses can be detected in less time than observation of CPE or plaques and they can detect viruses which do not produce CPE. Generally, virus replication can be detected in 2 to 5 days after inoculation but depends on the virus type [60]. In the United States, a study found that the use of ICC-PCR resulted in an increase in positive samples of surface water from 17.2% (5/29) by CPE to 93.1% (27/29) [61]. Studies conducted in South Korea [62] also reported greater isolation of naturally occurring enteric viruses by ICC-PCR and detection of enteric viruses in treated tap water that was previously negative by CPE. Similarly, a study conducted in New Zealand [45] reported greater numbers of viruses detected in surface waters using ICC-PCR than by qPCR.

Assays targeting viral messenger RNA for detection of human adenoviruses in environmental samples have been developed [47,63] but have not been widely applied in ambient waters. In addition, a molecular beacon-based real-time PCR assay has been applied to identify intact enteroviral particles combined with a reporter cell system to determine viral replication. The reporter assay depended upon fluorescence emitted by single-stranded dual-label antisense oligonucleotide probes (i.e., molecular bacons) upon binding to the specified target (e.g., mRNA) [64,65].

4.2. Direct Molecular Methods for Detecting Virus Infectivity

Various methods have been developed to determine the potential infectivity of enteric viruses directly by molecular methods. The potential application of these methods and their limitations have been reviewed [59,66,67]. The success of such methods depends on knowledge of the mechanism of inactivation of a particular virus and the site of action of a particular disinfectant [2,68]. Different virus types and strains may have different sites of action for a particular disinfectant. Thus, one method that may work for RNA viruses may not work for dsDNA viruses. In addition, complicating this approach is that some viruses, such as adenoviruses, rendered non-infectious by ultraviolet light can use host cell enzymes to repair DNA damages on their genome [31,68]. Inactivated viruses can still cause infection in cells through multiplicity reactivation [69]. This occurs when two viruses with their nucleic acids damaged in different regions of their genomes infect the same host cell resulting in a complete genome capable of replication.

Intercalating dyes, such as propidium monoazide (PMA) and ethidium monoazide (EMA) in conjunction with qPCR (PMA-RT-qPCR and PMA-qPCR for RNA or DNA viruses, respectively), have been used to determine the potential infectivity of enteric viruses in water [11,70,71]. Treatment of virus suspensions with platinum (IV) chloride (PtCl4) has also been applied to discriminate between potentially infectious and thermally inactivated enteric hepatitis viruses in environmental samples [12,72,73]. Two hypotheses underlay the use of intercalating dyes (i) a virus with a damaged capsid is not infectious, (ii) intercalating dyes can reach and bind the genomes to block specifically the amplification of defective particles [68]. However, the success of these methods depends on knowledge of the mechanism of inactivation of a particular virus and the site of action of a particular disinfectant [2,68,74].

Another qPCR-based framework has been described and used to estimate virus infectivity [75]. The framework quantifies damage to the entire genome based on the qPCR amplification of smaller sections, assuming single-hit inactivation and a Poisson distribution of damage. The framework offers the potential to monitor the infectivity of viruses that remain nonculturable or not easily grown in cell culture, such as norovirus.

5. Conclusions

Determining the concentration of infectious enteric viruses in water reuse systems will likely be problematic into the near future. No one cell culture system can detect all of the infectious viruses that may be present in an environmental sample. However, advances in molecular biology which allow us to detect the genome of viruses known to infect humans and animals in environmental samples have revealed that the number of viruses may be 100 to 1000 greater than that detected by cell culture [3,76]. This requires us to reassess what proportion of these viruses that are potentially infectious so that we can adequately assess the risk and design treatment systems to reduce the risk of exposure. The ratio of virus genome detected versus those detected by viral culture will be greatly affected by wastewater and wastewater treatment processes and will not be a constant value. For example, different disinfectants will affect different virus types differently (e.g., different sites of action on the viral capsid or genome), and the presence of resistant mutants or viruses capable of the use of host cell enzymes for repair (infectivity can be affected by choice of cell line). Perhaps the best approach at present is to use molecular methods to assess the presence of enteric viruses in untreated wastewater where most viruses can be expected to be infectious. This has been the approach for treatment requirements in water reuse applications for potable and non-potable purposes, including irrigation of crops traditionally consumed raw [4,77].

Another approach to consider is the use of a safety factor when estimating the true concentration of an infection virus in an environmental sample. This might be useful since no one method can detect all of the likely infectious virus present in environmental samples. When estimating risk from chemicals, it is common to take into consideration the uncertainty of using data on toxicity developed in animals to humans and the lack of data. Usually, safety factors of 10 to 100 are used to estimate acceptable levels of risk. While this may be useful for estimating levels of infectious virus in raw wastewaters, it becomes more problematic when dealing with treated wastewater and environmental waters. However, considering the factors outlined in this review affecting assays for enteric viruses that a safety factor of 10 would not be unreasonable.

Author Contributions: C.P.G. and W.Q.B. contributed equally to this manuscript.

Funding: This review was supported, in part by the United States Department of Agriculture-National Institute of Food and Agriculture. Grant number 20166800725064 that established CONSERVE A Center of Excellence at the Nexus of Sustainable Water Reuse, Water and Health.

Conflicts of Interest: The authors declare no conflict of interest.

References

1. Gibson, K.E. Viral pathogens in water: Occurrence, public health impact, and available control strategies. *Curr. Opin. Virol.* **2014**, *4*, 50–57. [CrossRef] [PubMed]
2. Gall, A.M.; Marinas, B.J.; Lu, Y.; Shisler, J.L. Waterborne Viruses: A Barrier to Safe Drinking Water. *PLoS Pathog.* **2015**, *11*, e1004867. [CrossRef] [PubMed]
3. Gerba, C.P.; Betancourt, W.Q.; Kitajima, M. How much reduction of virus is needed for recycled water: A continuous changing need for assessment? *Water Res* **2017**, *108*, 25–31. [CrossRef] [PubMed]
4. Anonymous. *Title 22 and 17 California Code of Regulations, July 16, 2015. Regulations Related to Recycled Water*; State Water Resources Control Board, DIvision of Drinking Water: Sacramento, CA, USA, 2015.
5. Kitajima, M.; Iker, B.C.; Pepper, I.L.; Gerba, C.P. Relative abundance and treatment reduction of viruses during wastewater treatment processes–identification of potential viral indicators. *Sci. Total Environ.* **2014**, *488–489*, 290–296. [CrossRef] [PubMed]
6. Bibby, K.; Peccia, J. Identification of viral pathogen diversity in sewage sludge by metagenome analysis. *Environ. Sci. Technol.* **2013**, *47*, 1945–1951. [CrossRef] [PubMed]
7. Ng, T.F.; Marine, R.; Wang, C.; Simmonds, P.; Kapusinszky, B.; Bodhidatta, L.; Oderinde, B.S.; Wommack, K.E.; Delwart, E. High variety of known and new RNA and DNA viruses of diverse origins in untreated sewage. *J. Virol.* **2012**, *86*, 12161–12175. [CrossRef] [PubMed]
8. Aw, T.G.; Howe, A.; Rose, J.B. Metagenomic approaches for direct and cell culture evaluation of the virological quality of wastewater. *J. Virol. Methods* **2014**, *210*, 15–21. [CrossRef] [PubMed]
9. Hamza, I.A.; Bibby, K. Critical issues in application of molecular methods to environmental virology. *J. Virol. Methods* **2019**, *266*, 11–24. [CrossRef] [PubMed]
10. Leifels, M.; Hamza, I.A.; Krieger, M.; Wilhelm, M.; Mackowiak, M.; Jurzik, L. From Lab to Lake—Evaluation of Current Molecular Methods for the Detection of Infectious Enteric Viruses in Complex Water Matrices in an Urban Area. *PLoS ONE* **2016**, *11*, e0167105. [CrossRef] [PubMed]

11. Leifels, M.; Shoults, D.; Wiedemeyer, A.; Ashbolt, N.; Sozzi, E.; Hagemeier, A.; Jurzik, L. Capsid Integrity qPCR-An Azo-Dye Based and Culture-Independent Approach to Estimate Adenovirus Infectivity after Disinfection and in the Aquatic Environment. *Water* **2019**, *11*, 1196. [CrossRef]
12. Randazzo, W.; Vasquez-Garcia, A.; Aznar, R.; Sanchez, G. Viability RT-qPCR to Distinguish Between HEV and HAV With Intact and Altered Capsids. *Front. Microbiol.* **2018**, *9*, 1973. [CrossRef] [PubMed]
13. Wigginton, K.R.; Kohn, T. Virus disinfection mechanisms: The role of virus composition, structure, and function. *Curr. Opin. Virol.* **2012**, *2*, 84–89. [CrossRef] [PubMed]
14. Wigginton, K.R.; Pecson, B.M.; Sigstam, T.; Bosshard, F.; Kohn, T. Virus inactivation mechanisms: Impact of disinfectants on virus function and structural integrity. *Environ. Sci. Technol.* **2012**, *46*, 12069–12078. [CrossRef] [PubMed]
15. Leland, D.S.; Ginocchio, C.C. Role of cell culture for virus detection in the age of technology. *Clin. Microbiol. Rev.* **2007**, *20*, 49–78. [CrossRef] [PubMed]
16. Condit, R.C. Principles of virology. In *Fields Virology*; Wolters Klewers/Lippincott Williams & Wilkins: Philadelphia, PA, USA, 2013; pp. 19–51.
17. Grabow, W.O.K.; ENupen, M. Comparison of Primary Kidney Cells with the BGM Cell Line for the Enumeration of Enteric Viruses in Water by Means of a Tube Dilution Technique. In *Viruses and Wastewater Treatment*; Goddard, M., Butler, M., Eds.; Elsevier: Pergamon, Turkey, 1981; pp. 253–256.
18. Schmidt, N.J.; Ho, H.H.; Riggs, J.L.; Lennette, E.H. Comparative sensitivity of various cell culture systems for isolation of viruses from wastewater and fecal samples. *Appl. Environ. Microbiol.* **1978**, *36*, 480–486. [PubMed]
19. Dahling, D.R.; Berg, G.; Berman, D. BGM, a continuous cell line more sensitive than primary rhesus and African green kidney cells for the recovery of viruses from water. *Health Lab. Sci.* **1974**, *11*, 275–282.
20. Dahling, D.R.; Wright, B.A. Optimization of the BGM cell line culture and viral assay procedures for monitoring viruses in the environment. *Appl. Environ. Microbiol.* **1986**, *51*, 790–812.
21. Morris, R. Detection of Enteroviruses: An Assessment of Ten Cell Lines. *Water Sci. Technol.* **1985**, *17*, 81–88. [CrossRef]
22. Benton, W.H.; Ward, R.L. Induction of cytopathogenicity in mammalian cell lines challenged with culturable enteric viruses and its enhancement by 5-iododeoxyuridine. *Appl. Environ. Microbiol.* **1982**, *43*, 861–868.
23. Pinto, R.M.; Gajardo, R.; Abad, F.X.; Bosch, A. Detection of fastidious infectious enteric viruses in water. *Environ. Sci. Technol.* **1995**, *29*, 2636–2638. [CrossRef]
24. Benton, W.H.; Hurst, C.J. Evaluation of mixed cell types and 5-iodo-2′-deoxyuridine treatment upon plaque assay titers of human enteric viruses. *Appl. Environ. Microbiol.* **1986**, *51*, 1036–1040. [PubMed]
25. Rodriguez, R.A.; Gundy, P.M.; Gerba, C.P. Comparison of BGM and PLC/PRC/5 cell lines for total culturable viral assay of treated sewage. *Appl. Environ. Microbiol.* **2008**, *74*, 2583–2587. [CrossRef] [PubMed]
26. Dahling, D.R. Detection and enumeration of enteric viruses in cell culture. *Crit. Rev. Environ. Control* **1991**, *21*, 237–263. [CrossRef]
27. Cao, Y.; Walen, K.H.; Schnurr, D. Coxsackievirus B-3 selection of virus resistant Buffalo green monkey kidney cells and chromosome analysis of parental and resistant cells. *Arch. Virol.* **1988**, *101*, 209–219. [CrossRef] [PubMed]
28. Payment, P.; Trudel, M. Influence of inoculum size, incubation temperature, and cell culture density on virus detection in environmental samples. *Can. J. Microbiol.* **1985**, *31*, 977–980. [CrossRef] [PubMed]
29. Guo, H.; Chu, X.; Hu, J. Effect of Host Cells on Low- and Medium-Pressure UV Inactivation of Adenoviruses. *Appl. Environ. Microbiol.* **2010**, *76*, 7068–7075. [CrossRef] [PubMed]
30. Cashdollar, J.L.; Huff, E.; Ryu, H.; Grimm, A.C. The influence of incubation time on adenovirus quantitation in A549 cells by most probable number. *J. Virol. Methods* **2016**, *237*, 200–203. [CrossRef] [PubMed]
31. Eischeid, A.C.; Meyer, J.N.; Linden, K.G. UV Disinfection of Adenoviruses: Molecular Indications of DNA Damage Efficiency. *Appl. Environ. Microbiol.* **2009**, *75*, 23–28. [CrossRef]
32. Croci, L.; Cozzi, L.; Stacchini, A.; De Medici, D.; Toti, L. A rapid tissue culture assay for the detection of okadaic acid and related compounds in mussels. *Toxicon* **1997**, *35*, 223–230. [CrossRef]
33. Matza-Porges, S.; Eisen, K.; Ibrahim, H.; Haberman, A.; Fridlender, B.; Joseph, G. A new antiviral screening method that simultaneously detects viral replication, cell viability, and cell toxicity. *J. Virol. Methods* **2014**, *208*, 138–143. [CrossRef]

34. Karri, V.; Kumar, V.; Ramos, D.; Oliveira, E.; Schuhmacher, M. An in vitro cytotoxic approach to assess the toxicity of heavy metals and their binary mixtures on hippocampal HT-22 cell line. *Toxicol. Lett.* **2018**, *282*, 25–36. [CrossRef] [PubMed]
35. Hejkal, T.W.; Gerba, C.P.; Rao, V.C. Reduction of cytotoxicity in virus concentrates from environmental samples. *Appl. Environ. Microbiol.* **1982**, *43*, 731–733. [PubMed]
36. Hurst, C.J.; Goyke, T. Reduction of interfering cytotoxicity associated with wastewater sludge concentrates assayed for indigenous enteric viruses. *Appl. Environ. Microbiol.* **1983**, *46*, 133–139. [PubMed]
37. Sedmak, G.; Bina, D.; Macdonald, J.; Couillard, L. Nine-Year Study of the Occurrence of Culturable Viruses in Source Water for Two Drinking Water Treatment Plants and the Influent and Effluent of a Wastewater Treatment Plant in Milwaukee, Wisconsin (August 1994 through July 2003). *Appl. Environ. Microbiol.* **2005**, *71*, 1042–1050. [CrossRef] [PubMed]
38. Abbaszadegan, M.; Stewart, P.; LeChevallier, M. A Strategy for Detection of Viruses in Groundwater by PCR. *Appl. Environ. Microbiol.* **1999**, *65*, 444–449. [PubMed]
39. Genoni, A.; Canducci, F.; Rossi, A.; Broccolo, F.; Chumakov, K.; Bono, G.; Salerno-Uriarte, J.; Salvatoni, A.; Pugliese, A.; Toniolo, A. Revealing enterovirus infection in chronic human disorders: An integrated diagnostic approach. *Sci. Rep.* **2017**, *7*, 5013. [CrossRef] [PubMed]
40. Spendlove, R.S.; Schaffer, F.L. Enzymatic enhancement of infectivity of reovirus. *J. Bacteriol.* **1965**, *89*, 597–602. [PubMed]
41. Mahalanabis, M.; Reynolds, K.A.; Pepper, I.L.; Gerba, C.P. Comparison of Multiple Passage Integrated Cell Culture-PCR and Cytopathogenic Effects in Cell Culture for the Assessment of Poliovirus Survival in Water. *Food Environ. Virol.* **2010**, *2*, 225–230. [CrossRef]
42. Racaniello, V.R. Picornaviridae: The viruses and their replication. In *Fields Virology*; Dmhp, C.J.K., Griffin, D.E., Lamb, R.A., Martin, M.A., Racaniello, V.R., Roizman, B., Eds.; Lippincott Williams & Wilkins: Philadelphia, PA, USA, 2013; pp. 453–489.
43. Ward, R.L.; Knowlton, D.R.; Pierce, M.J. Efficiency of human rotavirus propagation in cell culture. *J. Clin. Microbiol.* **1984**, *19*, 748–753. [PubMed]
44. Donia, D.; Bonanni, E.; Diaco, L.; Divizia, M. Statistical correlation between enterovirus genome copy numbers and infectious viral particles in wastewater samples. *Lett. Appl. Microbiol.* **2010**, *50*, 237–240. [CrossRef] [PubMed]
45. Dong, Y.; Kim, J.; Lewis, G.D. Evaluation of methodology for detection of human adenoviruses in wastewater, drinking water, stream water and recreational waters. *J. Appl. Microbiol.* **2010**, *108*, 800–809. [CrossRef] [PubMed]
46. Polston, P.M.; Rodriguez, R.A.; Seo, K.; Kim, M.; Ko, G.; Sobsey, M.D. Field evaluation of an improved cell line for the detection of human adenoviruses in environmental samples. *J. Virol. Methods* **2014**, *205*, 68–74. [CrossRef] [PubMed]
47. Rodriguez, R.A.; Polston, P.M.; Wu, M.J.; Wu, J.; Sobsey, M.D. An improved infectivity assay combining cell culture with real-time PCR for rapid quantification of human adenoviruses 41 and semi-quantification of human adenovirus in sewage. *Water Res.* **2013**, *47*, 3183–3191. [CrossRef] [PubMed]
48. Galasso, G.J.; Sharp, J.; Sharp, D.G. The influence of degree of aggregation and virus quality on the plaque titer of aggregated vaccinia virus. *J. Immunol.* **1964**, *92*, 870–878. [PubMed]
49. Gerba, C.P.; Betancourt, W.Q. Viral Aggregation: Impact on Virus Behavior in the Environment. *Environ. Sci. Technol.* **2017**, *51*, 7318–7325. [CrossRef] [PubMed]
50. Carducci, A.; Cantiani, L.; Moscatelli, R.; Casini, B.; Rovini, E.; Mazzoni, F.; Giuntini, A.; Verani, M. Interference between enterovirus and reovirus as a limiting factor in environmental virus detection. *Lett. Appl. Microbiol.* **2002**, *34*, 110–113. [CrossRef] [PubMed]
51. Fout, G.S.; Dahling, D.R.; Safferman, R.S. *USEPA Manual of Methods for Virology*; Environmental Protection Agency: Washington, DC, USA, 2001; Chapter 15.
52. Dahling, D.R. An improved filter elution and cell culture assay procedure for evaluating public groundwater systems for culturable enteroviruses. *Water Environ. Res.* **2002**, *74*, 564–568. [CrossRef]
53. Richards, G.P.; Weinheimer, D.A. Influence of adsorption time, rocking, and soluble proteins on the plaque assay of monodispersed poliovirus. *Appl. Environ. Microbiol.* **1985**, *49*, 744–748.
54. Betancourt, W.Q.; Abd-Elmaksoud, S.; Gerba, C.P. Efficiency of Reovirus Concentration from Water with Positively Charged Filters. *Food Environ. Virol.* **2018**, *10*, 209–211. [CrossRef]

55. Reynolds, K.A. Integrated cell culture/PCR for detection of enteric viruses in environmental samples. *Methods Mol. Biol.* **2004**, *268*, 69–78.
56. Moce-Llivina, L.; Lucena, F.; Jofre, J. Double-layer plaque assay for quantification of enteroviruses. *Appl. Environ. Microbiol.* **2004**, *70*, 2801–2805. [CrossRef] [PubMed]
57. Reynolds, K.A.; Gerba, C.P.; Abbaszadegan, M.; Pepper, L.L. ICC/PCR detection of enteroviruses and hepatitis A virus in environmental samples. *Can. J. Microbiol.* **2001**, *47*, 153–157. [CrossRef] [PubMed]
58. Reynolds, K.A.; Gerba, C.P.; Pepper, I.L. Detection of infectious enteroviruses by an integrated cell culture-PCR procedure. *Appl. Environ. Microbiol.* **1996**, *62*, 1424–1427. [PubMed]
59. Rodriguez, R.A.; Pepper, I.L.; Gerba, C.P. Application of PCR-based methods to assess the infectivity of enteric viruses in environmental samples. *Appl. Environ. Microbiol.* **2009**, *75*, 297–307. [CrossRef] [PubMed]
60. Greening, G.E.; Hewitt, J.; Lewis, G.D. Evaluation of integrated cell culture-PCR (C-PCR) for virological analysis of environmental samples. *J. Appl. Microbiol.* **2002**, *93*, 745–750. [CrossRef] [PubMed]
61. Chapron, C.D.; Ballester, N.A.; Fontaine, J.H.; Frades, C.N.; Margolin, A.B. Detection of astroviruses, enteroviruses, and adenovirus types 40 and 41 in surface waters collected and evaluated by the information collection rule and an integrated cell culture-nested PCR procedure. *Appl. Environ. Microbiol.* **2000**, *66*, 2520–2525. [CrossRef] [PubMed]
62. Lee, H.K.; Jeong, Y.S. Comparison of total culturable virus assay and multiplex integrated cell culture-PCR for reliability of waterborne virus detection. *Appl. Environ. Microbiol.* **2004**, *70*, 3632–3636. [CrossRef]
63. Ko, G.; Cromeans, T.L.; Sobsey, M.D. Detection of Infectious Adenovirus in Cell Culture by mRNA Reverse Transcription-PCR. *Appl. Environ. Microbiol.* **2003**, *69*, 7377–7384. [CrossRef] [PubMed]
64. Julian, T.R.; Schwab, K.J. Challenges in environmental detection of human viral pathogens. *Curr. Opin. Virol.* **2012**, *2*, 78–83. [CrossRef]
65. Hwang, Y.C.; Leong, O.M.; Chen, W.; Yates, M.V. Comparison of a reporter assay and immunomagnetic separation real-time reverse transcription-PCR for the detection of enteroviruses in seeded environmental water samples. *Appl. Environ. Microbiol.* **2007**, *73*, 2338–2340. [CrossRef]
66. Manuel, C.S.; Moore, M.D.; Jaykus, L.A. Predicting human norovirus infectivity - Recent advances and continued challenges. *Food Microbiol.* **2018**, *76*, 337–345. [CrossRef] [PubMed]
67. Girones, R.; Ferrus, M.A.; Alonso, J.L.; Rodriguez-Manzano, J.; Calgua, B.; Correa Ade, A.; Hundesa, A.; Carratala, A.; Bofill-Mas, S. Molecular detection of pathogens in water–the pros and cons of molecular techniques. *Water Res.* **2010**, *44*, 4325–4339. [CrossRef] [PubMed]
68. Prevost, B.; Goulet, M.; Lucas, F.S.; Joyeux, M.; Moulin, L.; Wurtzer, S. Viral persistence in surface and drinking water: Suitability of PCR pre-treatment with intercalating dyes. *Water Res.* **2016**, *91*, 68–76. [CrossRef] [PubMed]
69. McClain, M.E.; Spendlove, R.S. Multiplicity reactivation of reovirus particles after exposure to ultraviolet light. *J. Bacteriol.* **1966**, *92*, 1422–1429. [PubMed]
70. Parshionikar, S.; Laseke, I.; Fout, G.S. Use of propidium monoazide in reverse transcriptase PCR to distinguish between infectious and noninfectious enteric viruses in water samples. *Appl. Environ. Microbiol.* **2010**, *76*, 4318–4326. [CrossRef] [PubMed]
71. Karim, M.R.; Fout, G.S.; Johnson, C.H.; White, K.M.; Parshionikar, S.U. Propidium monoazide reverse transcriptase PCR and RT-qPCR for detecting infectious enterovirus and norovirus. *J. Virol. Methods* **2015**, *219*, 51–61. [CrossRef] [PubMed]
72. Randazzo, W.; Piqueras, J.; Rodriguez-Diaz, J.; Aznar, R.; Sanchez, G. Improving efficiency of viability-qPCR for selective detection of infectious HAV in food and water samples. *J. Appl. Microbiol.* **2018**, *124*, 958–964. [CrossRef]
73. Randazzo, W.; Piqueras, J.; Evtoski, Z.; Sastre, G.; Sancho, R.; Gonzalez, C.; Sanchez, G. Interlaboratory Comparative Study to Detect Potentially Infectious Human Enteric Viruses in Influent and Effluent Waters. *Food Environ. Virol.* **2019**. [CrossRef]
74. Rodríguez-Lázaro, D.; Kovač, K.; Diez-Valcarce, M.; Hernández, M.; D'Agostino, M.; Muscillo, M.; Cook, N.; Ruggeri, F.M.; Sellwood, J.; Nasser, A.; et al. Virus hazards from food, water and other contaminated environments. *FEMS Microbiol. Rev.* **2012**, *36*, 786–814. [CrossRef]
75. Pecson, B.M.; Ackermann, M.; Kohn, T. Framework for using quantitative PCR as a nonculture based method to estimate virus infectivity. *Environ. Sci. Technol.* **2011**, *45*, 2257–2263. [CrossRef]

76. Gerba, C.P.; Betancourt, W.Q.; Kitajima, M.; Rock, C.M. Reducing uncertainty in estimating virus reduction by advanced water treatment processes. *Water Res.* **2018**, *133*, 282–288. [CrossRef] [PubMed]
77. WHO. *WHO Guidelines for the Safe Use of Wastewater, Excreta and Greywater (Volume IV: Excreta and Greywater use in Agriculture)*; World Health Organization: Geneva, Switzerland, 2006; ISBN 92 4 154685 9.

pathogens

MDPI

Article

Detection of Pathogenic Viruses, Pathogen Indicators, and Fecal-Source Markers within Tanker Water and Their Sources in the Kathmandu Valley, Nepal

Bikash Malla [1], Rajani Ghaju Shrestha [2], Sarmila Tandukar [3], Dinesh Bhandari [4], Ocean Thakali [5], Jeevan B. Sherchand [4] and Eiji Haramoto [1,*]

1 Interdisciplinary Center for River Basin Environment, University of Yamanashi, 4-3-11 Takeda, Kofu, Yamanashi 400-8511, Japan; mallabikash@hotmail.com
2 Division of Sustainable Energy and Environmental Engineering, Osaka University, Suita, Osaka 565-0871, Japan; rajani_ghaju12@hotmail.com
3 Department of Natural, Biotic and Social Environment Engineering, University of Yamanashi, 4-3-11 Takeda, Kofu, Yamanashi 400-8511, Japan; sar1234tan@gmail.com
4 Institute of Medicine, Tribhuvan University Teaching Hospital, Kathmandu 1524, Nepal; me.dinesh43@gmail.com (D.B.); jeevanbsherchand@gmail.com (J.B.S.)
5 Environmental and Social System Science Course, University of Yamanashi, 4-3-11 Takeda, Kofu, Yamanashi 400-8511, Japan; othakali@gmail.com
* Correspondence: eharamoto@yamanashi.ac.jp; Tel.: +81-55-220-8725

Received: 21 May 2019; Accepted: 17 June 2019; Published: 19 June 2019

Abstract: Tanker water is used extensively for drinking as well as domestic purposes in the Kathmandu Valley of Nepal. This study aimed to investigate water quality in terms of microbial contamination and determine sources of fecal pollution within these waters. Thirty-one samples from 17 tanker filling stations (TFSs) and 30 water tanker (WT) samples were collected during the dry and wet seasons of 2016. *Escherichia coli* was detected in 52% of the 31 TFS samples and even more frequently in WT samples. Of the six pathogenic viruses tested, enteroviruses, noroviruses of genogroup II (NoVs-GII), human adenoviruses (HAdVs), and group A rotaviruses were detected using quantitative PCR (qPCR) at 10, five, four, and two TFSs, respectively, whereas Aichi virus 1 and NoVs-GI were not detected at any sites. Index viruses, such as pepper mild mottle virus and tobacco mosaic virus, were detected using qPCR in 77% and 95% out of 22 samples, respectively, all of which were positive for at least one of the tested pathogenic viruses. At least one of the four human-associated markers tested (i.e., BacHum, HAdVs, and JC and BK polyomaviruses) was detected using qPCR in 39% of TFS samples. Ruminant-associated markers were detected at three stations, and pig- and chicken-associated markers were found at one station each of the suburbs. These findings indicate that water supplied by TFSs is generally of poor quality and should be improved, and proper management of WTs should be implemented.

Keywords: fecal-source marker; index virus; microbial contamination; pathogenic virus; tanker water

1. Introduction

Kathmandu, the capital city of Nepal, faces a severe scarcity of water in terms of both quality and quantity [1–4]. Kathmandu Upatyaka Khanepani Limited (KUKL), the sole organization responsible for supplying piped water into the valley, can only supply 111 million liters per day (MLD) and 71 MLD in wet and dry seasons, respectively, while the actual demand approaches 377 MLD [4]. Therefore, to meet daily requirements for domestic water, households in the valley are compelled to employ alternative water sources [5]. Commonly used alternative water sources include groundwater (e.g., shallow dug and deep tube wells, and stone spouts), jar water, tanker water, and surface water sources, such as springs and rivers. Tanker water is a major component of the valley's water market [6], as is so in other countries, such as Bangladesh, Indonesia, Pakistan, the Philippines, and Thailand [7]. Water tankers play an important role in transporting large volumes of water abstracted from ground and surface sources to communities and households lacking the infrastructure or that are deprived of water sources [6,8–10]. The sources of tanker water in the valley range from surface water to shallow or deep borings, whereas the treatment procedures usually applied by TFSs vary from aeration, sedimentation and filtration (generally by pressurized sand filters), to use of bleaching powders [6]. The number of tanker water consumers has been gradually increasing and has increased rapidly following the Gorkha Earthquake of 2015 [5]. Currently, 22% of households are using tanker water, of which 18%, 60%, 97%, and 95% use it for drinking, cooking, bathing, and laundry, respectively [5].

A previous study [11] reported the detection of fecal indicator bacteria and pathogens as well as ruminant fecal markers in tanker water supplied to a household. A recent study showed that 77% of tanker water samples collected in the valley exceeded the Nepal Drinking Water Quality Standard guideline for total coliform count [12]. Such findings have indicated possible public health risks associated with using tanker water.

Viruses such as pepper mild mottle virus (PMMoV) and tobacco mosaic virus (TMV) have been proposed as potential indicators of pathogenic viruses [13]. Pathogenic viruses, including Aichi virus 1 (AiV-1), human adenoviruses (HAdVs), enteroviruses (EVs), noroviruses of genogroups I and II (NoVs-GII), and group A rotaviruses (RVAs), have been studied to estimate the concentration of pathogenic viruses in various water sources [13,14]. However, data regarding tanker water are limited. Thus, there is a need to investigate microbial contamination and sources of fecal pollution in TFS samples and water distributed by WTs.

Prevention of potential disease outbreaks can be achieved by identifying sources of fecal contamination and formulating appropriate pollution mitigation strategies. Sources of fecal contamination can be identified by the application of a technique called microbial source tracking (MST), which accurately and reliably identifies the hosts responsible for fecal pollution [15,16]. Host-associated *Bacteroidales* assays—BacHum (human-associated) [17], BacR (ruminant-associated) [18], and Pig2Bac (pig-associated) [19] and mitochondrial DNA (mtDNA) markers (bovine-, dog-, and pig-associated) [20,21], as well as viral markers specific for humans (HAdVs) [22], JC and BK polyomaviruses (JCPyVs and BKPyVs) [23], chicken (chicken parvoviruses (ChkPVs) [24], and pig (porcine adenoviruses (PoAdVs) [25])—are commonly used for source tracking.

Based on this background, the current study aimed to assess the prevalence and abundance of pathogenic viruses and indicators of pathogens in order to identify sources of fecal contamination in TFSs and WT samples in the Kathmandu Valley.

2. Results

2.1. Detection of Fecal Indicator Bacteria and Index Viruses

Table 1 shows the positive ratios and concentration ranges of fecal indicator bacteria and index viruses (PMMoV and TMV) within water samples from TFSs and WTs. *Escherichia coli* and total coliforms were detected in 52% and 87% of 31 TFS samples, respectively, and were more frequent in WT samples. The mean concentration of *E. coli* in WT samples was 0.37 log greater than that in TFS samples, although the difference was not significant (independent *t*-test; $p > 0.05$). PMMoV and TMV were detected in 71% and 90% out of 31 TFS samples, respectively, whereas in WT samples, PMMoV and TMV were detected in 73% and 97% out of 30 samples, respectively. Of the 22 samples that were positive for at least one pathogenic virus, PMMoV and TMV were detected in 77% and 95% of samples, respectively. The *E. coli* concentrations were 0.0–4.0 and 0.0–3.5 log most probable number (MPN)/100 mL in TFSs and WT samples, respectively. Similarly, out of the two index viruses tested, TMV was detected with the highest concentration (6.3 log copies/L) in WT samples, whereas PMMoV was detected with the lowest concentration (1.7 log copies/L) in TFS samples. *E. coli* was detected in

44% (7/16) and 60% (9/15) of TFS samples during the dry and wet seasons, respectively, whereas it was detected in 65% (11/17) and 77% (10/13) of WT samples during the dry and wet seasons, respectively. Although the difference was not significant, the mean concentration of *E. coli* in WT samples during the wet season was 0.78 log greater than that within the dry season (independent *t*-test; $p > 0.05$).

Figure 1 shows the *E. coli* concentration of water samples in the corresponding TFSs and WTs (27 pairs). In most cases, the *E. coli* concentration of WT samples was greater than that of corresponding TFS samples, although the mean concentrations did not differ significantly between WT (0.8 ± 1.6 log MPN/100 mL) and TFS samples (0.5 ± 1.8 log MPN/100 mL) (paired *t*-test, $p > 0.05$). Forty-six percent (6/13) of *E. coli*-negative TFS samples were positive for *E. coli* in the corresponding WT samples.

Chlorine is a widely used disinfectant employed within water treatment procedures in the valley. We examined the relationship between the *E. coli*-positive ratio and the concentrations of free and combined chlorine within TFS samples. Figure 2 shows the positive ratios of *E. coli* in water samples from TFSs in different categories of free (Figure 2a) and combined (Figure 2b) chlorine concentrations. The positive ratios of *E. coli* gradually decreased with an increase in free and combined chlorine concentrations, except for the category of 0.00–0.05 mg/L free chlorine. The concentration of total chlorine in this category was 0.01–0.59 mg/L. When water samples were divided into three categories based on total chlorine concentration, the positive ratios of *E. coli* were 60% (6/10), 64% (7/11), and 30% (3/10) for 0.01–0.04, 0.05–0.34, and 0.35–1.42 mg/L of chlorine concentration, respectively.

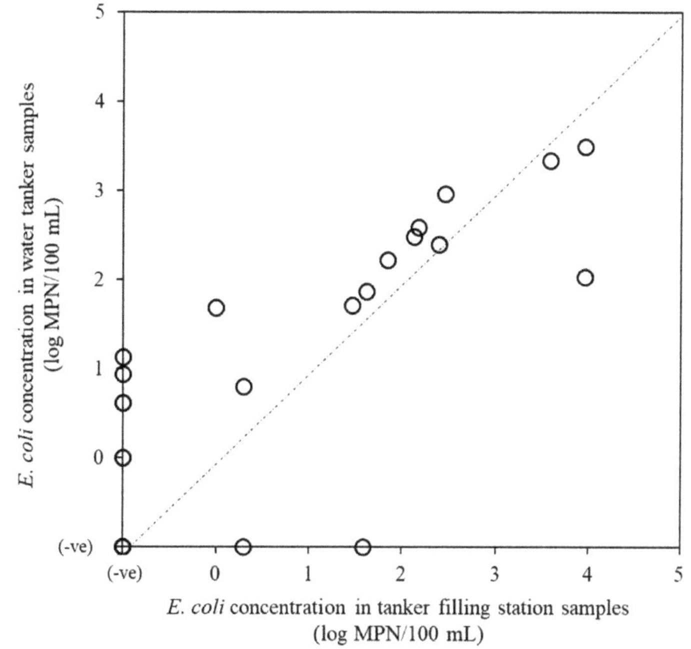

Figure 1. *E. coli* concentrations in tanker filling station and water tanker samples.

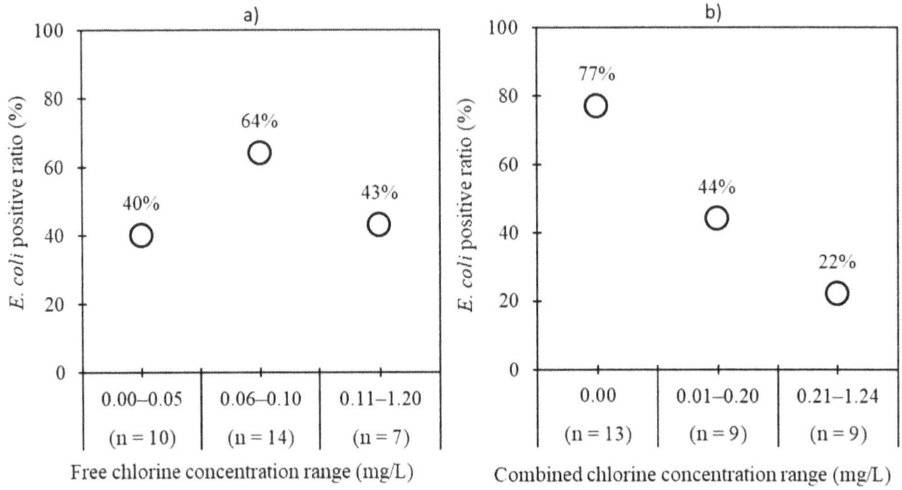

Figure 2. *E. coli* concentrations plotted against (**a**) free chlorine concentration categories and (**b**) combined chlorine concentration categories in tanker filling station samples.

2.2. Detection of Pathogenic Viruses

Table 2 shows the results of testing for six pathogenic viruses—AiV-1, EVs, HAdVs, NoVs-GI and GII, and RVAs—analyzed for TFS and WT samples. Of the 17 TFSs, EVs, NoVs-GII, HAdVs, and RVAs were detected at 10, five four, and two TFSs, respectively. Between two and four pathogenic viruses were detected at six TFSs. Among all the pathogenic viruses tested, EVs were the most prevalent viruses in TFS samples, with a positive ratio of 35% (11/31), followed by NoVs-GII (23%, 7/31), HAdVs (13%, 4/31), and RVAs (6%, 2/31). On the other hand, NoVs-GII were most frequently detected in WT samples (20%, 6/30), followed by EVs (13%, 4/30), RVAs (10%, 3/30), and HAdVs (7%, 2/30). The detection frequency of EVs was significantly higher in TFS samples (35%, 11/31) than in WT samples (13%, 4/30) (χ^2-test; $p < 0.05$). However, no significant differences in the detection frequencies of NoVs-GII (χ^2-test; $p > 0.05$), HAdVs, and RVAs (fisher exact-test; $p > 0.05$) between TFS and WT samples were observed. At least one pathogenic virus was detected in 45% (14/31) of TFS samples and 27% (8/30) of WT samples. Furthermore, NoVs-GII were detected at two TFSs continuously during both seasons. However, AiV-1 and NoVs-GI were undetected in any of the sampled TFSs and WTs.

Table 1. Positive ratios and concentrations of fecal indicator bacteria and index viruses in tanker filling station and water tanker samples.

Water Sample	No. of Tested Samples	Fecal Indicator Bacteria				Index Viruses			
		E. coli		Total Coliforms		PMMoV		TMV	
		No. of Positive Samples (%)	Concentration [a] (log MPN [b]/100 mL)	No. of Positive Samples (%)	Concentration [a] (log MPN [b]/100 mL)	No. of Positive Samples (%)	Concentration [a] (log copies/L)	No. of Positive Samples (%)	Concentration [a] (log copies/L)
Tanker filling station	31	16 (52)	0.0–4.0	27 (87)	0.0–5.4	22 (71)	1.7–4.7	28 (90)	2.7–6.0
Water tanker	30	21 (70)	0.0–3.5	27 (90)	1.0–4.8	22 (73)	2.1–4.9	29 (97)	2.8–6.3
Total	61	37 (61)		54 (89)		44 (72)		57 (93)	

[a] Range of concentrations among positive samples. [b] MPN, most probable number.

Table 2. Positive ratios and concentrations of pathogenic viruses in tanker filling station and water tanker samples.

Water Sample	No. of Tested Samples	AiV-1		EVs		HAdVs		NoVs-GI		NoVs-GII		RVAs		At Least One Pathogen Detected
		No. of Positive Samples (%)	Conc. [a] (log copies/L)	No. of Positive Samples (%)	Conc. [a] (log copies/L)	No. of Positive Samples (%)	Conc. [a] (log copies/L)	No. of Positive Samples (%)	Conc. [a] (log copies/L)	No. of Positive Samples (%)	Conc. [a] (log copies/L)	No. of Positive Samples (%)	Conc. [a] (log copies/L)	No. of Positive Samples (%)
Tanker filling station	31	0 (0)	NA	11 (35)	2.7–4.6	4 (13)	3.6–4.9	0 (0)	NA	7 (23)	2.0–3.9	2 (6)	3.3–3.7	14 (45)
Water tanker	30	0 (0)	NA	4 (13)	3.1–4.6	2 (7)	4.3–5.0	0 (0)	NA	6 (20)	1.8–4.5	3 (10)	2.8–3.4	8 (27)
Total	61	0 (0)		15 (25)		6 (10)		0 (0)		13 (21)		5 (8)		22 (36)

[a] Range of concentrations among positive samples; NA, not applicable.

2.3. Detection of Host-Associated Fecal Markers

Microbial source tracking was conducted for TFS samples using previously validated host-associated *Bacteroidales* [26], mtDNA, and viral markers. Table 3 shows the results of the detection of fecal markers in the TFS samples. The frequency of at least one human-associated marker (39%, 12/31) detection was significantly higher than ruminant-associated marker (14%, 3/22) (χ^2-test; $p < 0.05$). Chicken- and pig-associated markers were detected in 3% (1/31) and 5% (1/22) of TFS samples, respectively. Dog-associated markers were not detected in any of the TFS samples. At least one human- and ruminant-associated markers were detected at 10 and 3 out of 17 TFSs tested, respectively. Human- and animal-mixed fecal contamination was observed at two TFSs. For one TFS, contaminations from all the tested hosts were judged, with the exception of dog. Animal-associated fecal markers were detected at three TFSs, all of which were located in the peri-urban area where agriculture and livestock farming are common. At least one pathogenic virus was detected in 69% (9/13) and 33% (6/18) of samples that tested positive and negative for fecal markers, respectively. At least one fecal marker was detected at nine (75%) out of 12 TFSs within which pathogenic viruses were detected. In addition, human-associated fecal markers were continuously detected at two TFSs during both seasons.

Table 3. Detection of fecal-source markers in tanker filling station samples.

Fecal Markers		Detection % (No. of Positive Samples/No. of Tested Samples)	Concentration [d] (log copies/L)
Human-	BacHum [a]	5 (1/22)	6.3
	HAdVs [b]	13 (4/31)	3.6–4.9
	BKPyVs [b]	29 (9/31)	4.9–5.7
	JCPyVs [b]	10 (3/31)	5.0–5.9
	At least one human marker	39 (12/31)	3.6–6.3
Ruminant-	BacR [a]	14 (3/22)	5.4–5.9
	Bovine mtDNA [c]	0 (0/22)	NA [e]
Pig-	Pig2Bac [a]	5 (1/22)	6.1
	PoAdVs [b]	0 (0/31)	NA
	Swine mtDNA [c]	0 (0/22)	NA
Dog-	Dog mtDNA [c]	0 (0/22)	NA
Chicken-	ChkPVs [b]	3 (1/31)	3.4

[a] *Bacteroidales* marker; [b] Viral marker; [c] Mitochondrial DNA marker; [d] Range of concentrations among positive samples; [e] NA, not applicable.

3. Discussion

Fifty-two percent (16/31) of TFS samples were contaminated with *E. coli*, indicating poor performance of the treatment plants. *E. coli* detection in 70% (21/30) of WT samples with concentrations higher than the World Health Organization (WHO) guideline values for drinking water (<1 MPN/100 mL) indicated the unsuitability of this tanker water for drinking purposes [27]. When the relationship between *E. coli* detection and free or combined chlorine concentrations was examined, there was a decreasing trend in the positive ratios of *E. coli* as the concentrations of free and combined chlorine increased. However, there was a low positive ratio of *E. coli* in the category 0.00–0.05 mg/L of free chlorine, which could be due to the presence of combined chlorine. This result suggested that chlorine application could be a useful measure for lowering the concentration of *E. coli* in WTs. Although the difference was not significant, the concentrations of *E. coli* in WT samples were higher compared with their corresponding TFS samples. *E. coli* was detected in 46% (6/13) of WT samples that were negative for the corresponding TFSs. These results indicated that tankers are not disinfected and/or cleaned regularly. A similar result was obtained in Lebanon, where eight tankers had higher concentrations of fecal coliforms than their water sources [28].

High positive ratios for the potential indicators of pathogenic viruses, PMMoV and TMV, in TFS and WT samples indicated that other water-transmitted viral pathogens, such as astroviruses and hepatitis A and E viruses, could be present, for which testing was not performed in this study. Group A rotaviruses, which are the major causative agent of gastroenteritis in Nepal [29–31], were detected in 10% (3/30) of WT samples. Previous studies have reported the detection of pathogenic viruses—such as AiV-1, EVs, HAdVs, NoVs-GI, NoVs-GII, and RVAs—in groundwater and river water in the valley, which are the major sources of tanker water [1,13,14,32,33]. A tap water sample supplied by a tanker in the valley was found to be contaminated with pathogens, including HAdVs and *Vibrio cholerae*, further indicating the unsuitability of tanker water for drinking purposes [11]. In addition, NoVs-GI and HAdVs were also detected in two and one samples, respectively, out of five water tankers sampled in the valley, and enteric viruses were found to be responsible for gastroenteritis in children suffering from diarrhea [33]. A previous study reported a high risk of diarrheal infections for consumers of raw vegetables washed with tanker water or other water sources in the valley [34]. High positive ratios of fecal indicator bacteria and pathogenic viruses in TFS samples show that the employed treatment systems were not sufficient to eliminate the pathogens tested.

When the possible sources of such pathogenic viruses and fecal indicator bacteria in these water samples were analyzed by an MST technique, 39% (12/31) and 14% (3/22) of water samples were judged to be contaminated with human and ruminant feces, respectively. The detection of ruminant fecal markers has been previously reported in tanker water [11]. This could be due to the use of groundwater and surface water by the TFSs, in which human and animal fecal contaminations have been reported [11,35,36]. A previous study reported the possible transmission of enteric viruses from feces to children consuming water from sources contaminated by these viruses [33]. The detection of pathogenic viruses and fecal markers in the same sample indicated that these viruses might have originated from the feces of humans and animals. The detection of the animal fecal markers, mostly in samples originating from the peri-urban areas of the valley, could be due to the land use pattern of those areas where agriculture and farming are commonly practiced [35]. In Cambodia, animals were found to be responsible for the fecal pollution of water sources in agricultural areas [37], and livestock ownership is significantly associated with water contamination in Ghana and Bangladesh [38]. These results indicate a high risk to public health, which requires immediate action for control and prevention of possible disease outbreaks.

Groundwater, a major source for tanker water in Nepal [6,9], is contaminated by human and animal feces [26,35]. Despite an effort to ban on the implementation of deep tube wells within a 200 m distance of riverbanks, some TFSs are still found near riverbanks. Mixing of river water with nearby groundwater has been previously reported [39]. These reasons may contribute to the poor microbial quality of tanker water. This study showed that an increase in the concentrations of free and combined chlorine was associated with decreased concentrations of *E. coli* in WT samples, suggesting that chlorine application could be one of the measures used to lower the concentration of *E. coli* in WTs.

In conclusion, this study reports that the water supplied to the TFSs and WTs to the public are contaminated with fecal indicator bacteria and pathogenic viruses. This study also highlighted the use of host-associated *Bacteroidales*, mtDNA, and viral genetic markers to identify the sources of fecal pollution. The major source of microbial contamination was judged to be human feces, indicating that better infrastructure and management practices should be implemented. The increased microbial contamination present in WTs compared with that of TFS samples suggests the importance of regular cleaning and disinfection of the WTs.

4. Materials and Methods

4.1. Collection of Water Samples

Altogether, 31 TFS water samples were collected from 17 TFSs during the dry (March; n = 16) and wet (August; n = 15) seasons of 2016, and from 30 WTs during the dry (n = 17) and wet (n = 13) seasons of the same year. The water supplied by the tanker water treatment plants or TFSs to the tankers or the vehicles that carry water are referred to as TFS samples, and the water distributed by these vehicles to the public are referred to as WT samples. Water samples were collected in two 100 mL and five 1 L plastic bottles, which were washed with pure water prior to autoclaving, for each of the TFS and WT samples. Chlorine concentrations of WT samples were measured using a portable water analyzer colorimeter (HACH, Loveland, Co, USA). All samples were stored cold, transported to the laboratory, kept at 4 °C, and processed within 4 h.

4.2. Detection of Total Coliforms and *E. coli*

Total coliforms and *E. coli* were determined by the MPN method using a Colilert reagent (IDEXX Laboratories, Westbrook, CA, USA), as described previously [14,40].

4.3. Concentration and Extraction of Bacterial, mtDNA, and Viral Markers and Viruses

Bacterial and mtDNA were extracted using a CicaGeneus DNA Extraction Reagent (Kanto Chemical, Tokyo, Japan), as previously described [26,35]. Briefly, 100 mL of a water sample was filtered using a disposable filter unit preset with a nitrocellulose membrane (diameter, 47 mm; pore size, 0.22 μm; Nalgene, Tokyo, Japan). The filter membrane was transferred into a 50 mL tube and 5 mL of Tris–EDTA buffer (pH 7.4) was added. The resuspended sample was processed after repeated shaking and mixing by vortexing. A final volume of 300 μL of DNA extract was obtained by processing 160 μL of the resuspended sample with 20 μL of Buffer A and 200 μL of Buffer B.

An electronegative membrane-vortex method [41] was used as described previously with some modifications for virus concentration of the water samples [13,14,36]. Briefly, for the concentration step, 50 mL of 2.5 mol/L $MgCl_2$ was added to the 5 L water sample and filtered using a mixed cellulose-ester membrane (pore size, 0.8 μm; diameter, 90 mm; Merck Millipore, Billerica, MA, USA). Filter membrane was removed from the filter holder and vigorous vortexing of the membrane was performed with elution buffer in a 50 mL plastic tube to recover an eluate (~15 mL), as mentioned previously [13,14]. Subsequently, the eluate was centrifuged at 2000 × *g* for 10 min at 4 °C, followed by filtration of supernatant using a disposable membrane filter unit (pore size, 0.45 μm; diameter, 25 mm; Advantec, Tokyo, Japan). Finally, the filtrate was further concentrated using a Centriprep YM-50 ultrafiltration device (Merck Millipore) to obtain a virus concentrate, following the manufacturer's protocol. Viral DNA was extracted using a QIAamp DNA Mini Kit (QIAGEN, Hilden, Germany) from 200 μL of viral concentrate to obtain 200 μL of DNA extract. Similarly, a QIAamp Viral RNA Mini Kit (QIAGEN) was used to obtain a 60 μL RNA extract from 140 μL of viral concentrate, following the manufacturer's protocol. Both DNA and RNA extractions were performed using a QIAcube automated platform (QIAGEN). Thirty microliters of viral RNA was subjected to reverse transcription using a High-Capacity cDNA Reverse Transcription Kit (Applied Biosystems, Foster City, CA, USA) to obtain 60 μL of cDNA.

4.4. Detection of Viruses and Fecal Markers

The effect of qPCR inhibition was evaluated in this study as recommended elsewhere [42]. Porcine teschovirus (PoTeVs), as a control, was inoculated into DNA extract and recovered by qPCR. For quantitative PCR (qPCR), 2.5 μL of template DNA/cDNA was added to a mixture of 22.5 μL containing 12.5 μL Probe qPCR Mix (Takara Bio, Kusatsu, Japan), 7.0 μL PCR-grade water, 1.0 μL each of 10 pmol/μL forward and reverse primers, and 1.0 μL of the 5 pmol/μL TaqMan (MGB) probe. Table 4 shows the sequences of primers and probes used in this study. For the quantification of genomes, a Thermal Cycler Dice Real Time System TP800 (Takara Bio) was used. The thermal cycle conditions for all the tested assays (BacHum [17], BacR [18], Pig2Bac [19], Bovine- and Swine-mtDNA [20], Dog-mtDNA [21], AiV-1 [43], BKPyVs and JCPyVs [44], ChkPVs [24], and PoAdVs [25]) were as follows: 95 °C for 30 s, followed by 45 cycles at 95 °C for 5 s, and 60 °C for 30 s, except for EVs [45,46], PMMoV [47,48], RVAs [49], and TMV [50] (60 °C for 60 s), HAdVs [51], NoVs-GI, and NoVs-GII [52] (58 °C for 30 s), and PoTeVs [53] (56 °C for 30 s). For the determination of the genome copy number of each virus, a standard curve was plotted using six 10-fold serial dilutions of artificially synthesized plasmid DNA containing the amplification region. The amplification efficiencies of standard curves ranged from 78% to 123%. The calculated mean efficiency of process control was 141 ± 32% (n = 30), suggesting that there was no inhibition during qPCR.

In all qPCR runs, unknown and standard samples and negative controls were run in duplicate. A negative control was included in every run. The sample was judged positive if the respective marker was detected in at least one of the two wells with the threshold cycle value of ≤40.

4.5. Statistical Analysis

An independent *t*-test was used for the comparison of the *E. coli* concentrations between WT and TFS samples and for comparing the concentrations of *E. coli* in WT samples between dry and wet seasons. In addition, a paired *t*-test was used to compare the concentrations of *E. coli* between WT and corresponding TFS water samples. The detection frequencies of pathogenic viruses in TFS and WT samples were compared using χ^2 and Fisher Exact tests. Similarly, the χ^2 test was used for the comparison of the detection frequencies of human- and ruminant-associated markers in TFS samples. For negative samples, the one-tenth value of the limit of detection (1 MPN/100 mL for *E. coli*) was used. For statistical analyses, SPSS version 23 (IBM Corporation, Armonk, USA) was used, and values were considered significant at $p < 0.05$.

Table 4. Primer and probe sequences used in this study.

Assay	Primer/Probe	Sequence (5′–3′)	Product Length (bp)	Reference
AiV-1	Forward primer Reverse primer TaqMan MGB probe	GTCTCCACHGACACYAAYTGGAC GTTGTACATRGCAGCCCAGG FAM-TTYTCCTTYGTGCGTGC-MGB-NFQ	108–111	[43]
BacHum	Forward primer Reverse primer TaqMan probe	TGAGTTCACATGTCCGCATGA CGTTACCCCGCCTACTATCTAATG FAM-TCCGGTAGACGATGGGGATGCGTT-TAMRA	82	[17]
BacR	Forward primer Reverse primer TaqMan MGB probe	GCGTATCCAACCTTCCCG CATCCCCATCCGTTACCG FAM-CTTCCGAAAGGGAGATT-MGB-NFQ	118	[18]
BKPyVs	Forward primer Reverse primer TaqMan probe	GGCTGAAGTATCTGAGACTTGGG GAAACTGAAGACTCTGGACATGGA FAM-CAAGCACTGAATCCCAATCACAATGCTC-TAMRA	78	[44]
Bovine-mtDNA	Forward primer Reverse primer TaqMan probe	CAGCAGCCCTACAAGCAATGT GAGGCCAAATTGGGCGGATTAT FAM-CATCGGCGACATTGGTTTCATTTTAG-TAMRA	191	[20]
ChkPVs	Forward primer Reverse primer TaqMan probe	AGTCCACGAGATTGGCAACA GCAGGTTAAAGATTTTCACG FAM-AATTATTCGAGATGGCGCCCACG-TAMRA	82	[24]
Dog-mtDNA	Forward primer Reverse primer TaqMan probe	GGCATGCCTTTCCTTACAGGATTC GGGATGTGGCAACGAGTGTAATTATG FAM-TCATCGAGTCCGCTAACACGTCGAAT-TAMRA	109	[21]
EVs	Forward primer Reverse primer TaqMan probe	CCTCCGGCCCCTGAATG ACCGGATGGCCAATCCAA FAM-CCGACTACTTTGGGTGTCCGTGTTTC-TAMRA	195	[45] [46]
HAdVs	Forward primer Reverse primer TaqMan probe	GCCACGGTGGGGTTTCTAAACTT GCCCCAGTGGTCTTACATGCACATC FAM-TGCACCAGACCCGGGCTCAGGTACTCCGA-TAMRA	132	[51]
JCPyVs	Forward primer Reverse primer TaqMan probe	GGAAAGTCTTTAGGGTCTTCTACCTTT ATGTTTGCCAGTGATGATGAAAA FAM-GATCCCAACACTCTACCCCACCTAAAAAGA-TAMRA	89	[44]
NoVs-GI	Forward primer Reverse primer TaqMan probe	CGYTGGATGCGNTTYCATGA CTTAGACGCCATCATCATTYAC FAM-AGATYGCGATCYCCTGTCCA-TAMRA	85	[52]
NoVs-GII	Forward primer Reverse primer TaqMan probe	CARGARBCNATGTTYAGRTGGATGAG TCGACGCCATCTTCATTCACA FAM-TGGGAGGGCGATCGCAATCT-TAMRA	98	[52]

Table 4. *Cont.*

Assay	Primer/Probe	Sequence (5′–3′)	Product Length (bp)	Reference
Pig2Bac	Forward primer Reverse primer TaqMan MGB probe	GCATGAATTTAGCTTGCTAAATTTGAT ACCTCATACGGTATTAATCCGC FAM-TCCACGGGATAGCC-MGB-NFQ	117	[19]
PMMoV	Forward primer Reverse primer TaqMan MGB probe	GAGTGGTTTGACCTTAACGTTTGA TTGTCGGTTGCAATGCAAGT FAM-CCTACCGAAGCAAATG-MGB-NFQ	68	[47] [48] [47]
PoAdVs	Forward primer Reverse primer TaqMan MGB probe	AACGGCCGCTACTGCAAG AGCAGCAGGCTCTTGAGG FAM-CACATCCAGGTGCCGC-MGB-NFQ	68	[25]
PoTeVs	Forward primer Reverse primer TaqMan probe	CACCAGCGTGGAGTTCCTGTA AGCCCGCGACCCTGTCA FAM-TGCAGGACTGGACTTG-TAMRA	66	[53]
RVAs	Forward primer Reverse primer TaqMan probe	CAGTGGTTGATGCTCAAGATGGA TCATTGTAATCATATTGAATACCA FAM-ACAACTGCAGCTTCAAAAGAAGWGT-TAMRA	131	[49]
Swine-mtDNA	Forward primer Reverse primer TaqMan probe	ACAGCTGCACTACAAGCAATGC GGATGTAGTCCGAATTGAGCTGATTAT FAM-CATCGGAGACATTGGATTTGTCCTAT-TAMRA	197	[20]
TMV	Forward primer Reverse primer TaqMan probe	CAAGCTGGAACTGTCGTTCA CGGGTCTAAYACCGCATTGT FAM-CAGTGAGGTGTGGAAACCTTCACCACA-TAMRA	120	[50]

FAM, 6-carboxyfluorescein; MGB, minor groove binder; NFQ, nonfluorescent quencher; TAMRA, 5-carboxytetramethylrhodamine.

Author Contributions: B.M. conceived the design of the study, processed the samples, analyzed the results, and prepared a draft of the manuscript. R.G.S, S.T., D.B., and O.T. processed the samples. J.B.S. conceived the design of the study. E.H. conceived the design of the study, checked the analyzed results, and corrected the draft of the manuscript.

Funding: This study was supported by the Science and Technology Research Partnership for Sustainable Development (SATREPS) project entitled 'Hydro-microbiological approach for the water security in Kathmandu Valley, Nepal', and the Japan Society for the Promotion of Science (JSPS) through a Grant-in-Aid for Scientific Research (B) (grant number JP17H03332) and the Fund for the Promotion of Joint International Research (Fostering Joint International Research (B)) (grant number JP18KK0297).

Conflicts of Interest: The authors declare no conflict of interest.

References

1. Haramoto, E.; Yamada, K.; Nishida, K. Prevalence of protozoa, viruses, coliphages and indicator bacteria in groundwater and river water in the Kathmandu Valley, Nepal. *Trans. R. Soc. Trop. Med. Hyg.* **2011**, *105*, 711–716. [CrossRef] [PubMed]
2. Shrestha, S.; Malla, S.S.; Aihara, Y.; Kondo, N.; Nishida, K. Water Quality at supply source and point of use in Kathmandu Valley. *JWET* **2013**, *11*, 331–340. [CrossRef]
3. Guragai, B.; Takizawa, S.; Hashimoto, T.; Oguma, K. Effects of inequality of supply hours on consumers' coping strategies and perceptions of intermittent water supply in Kathmandu Valley, Nepal. *Sci. Total Environ.* **2017**, *599–600*, 431–441. [CrossRef] [PubMed]
4. KUKL. *KUKL 9th Annual Report*; Kathmandu Upatyaka Khanepani Limited: Kathmandu, Nepal, 2017.
5. Shrestha, S.; Aihara, Y.; Bhattarai, A.P.; Bista, N.; Rajbhandari, S.; Kondo, N.; Kazama, F.; Nishida, K.; Shindo, J. Dynamics of domestic water consumption in the urban area of the Kathmandu Valley: Situation analysis pre and post 2015 Gorkha Earthquake. *Water* **2017**, *9*, 222. [CrossRef]
6. Dongol, R.; Kansakar, L.K.; Bajimaya, S.; Maharjan, S.; Shrestha, D. Overview of water markets in the Kathmandu Valley. In *Kathmandu Valley Groundwater Outlook*; Shrestha, S., Pradhananga, D., Pandey, V.P., Eds.; SEN: Kathmandu, Nepal; CREEW: Kathmandu, Nepal; ICRE-UY: Yamanashi, Japan; Asian Institute of Technology: Pathum Thani, Thailand, 2012; pp. 100–111. ISBN 978-9937-2-4442-8.
7. Kejjlen, M.; Mcgranahan, G. *Informal Water Vendors and the Urban Poor Human Settlements Discussion Paper Series*; International Institute for Environment and Development: London, UK, 2006.

8. World Health Organization (WHO); United Nations Children's Fund (UNICEF). *Core Questions on Drinking-water and Sanitation for Household Surveys*; WHO: Geneva, Switzerland; UNICEF: Geneva, Switzerland, 2006; p. 125.
9. Pandey, V.P.; Chapagain, S.K.; Shrestha, D.; Shrestha, S.; Kazama, F. Groundwater Markets for Domestic Water Use in Kathmandu Valley: An Analysis of Its Characteristics, Impacts and Regulations. Available online: https://www.academia.edu/11198401/Groundwater_markets_for_domestic_water_use_in_Kathmandu_Valley_an_analysis_of_its_characteristics_impacts_and_regulations (accessed on 13 May 2019).
10. Shrestha, D. State and Services of Private Water Tanker Operation in Kathmandu. Unpublished M.Sc. Thesis, Nepal Engineering College, Pokhara University, Changunarayan, Nepal, 2011.
11. Haramoto, E. Detection of waterborne protozoa, viruses, and bacteria in groundwater and other water samples in the Kathmandu Valley, Nepal. *IOP Conf. Ser. Earth Environ. Sci.* **2018**, *120*, 012004. [CrossRef]
12. Maharjan, S.; Joshi, T.P.; Shrestha, S.J. Poor quality of treated water in Kathmandu: Comparison with Nepal drinking water quality standards. *Tribhuvan Univ. J. Microbiol.* **2018**, *5*, 83–88. [CrossRef]
13. Shrestha, S.; Shrestha, S.; Shindo, J.; Sherchand, J.B.; Haramoto, E. Virological quality of irrigation water sources and pepper mild mottle virus and tobacco mosaic virus as index of pathogenic virus contamination level. *Food Environ. Virol.* **2018**, *10*, 107–120. [CrossRef] [PubMed]
14. Tandukar, S.; Sherchand, J.B.; Bhandari, D.; Sherchan, S.; Malla, B.; Ghaju Shrestha, R.; Haramoto, E. Presence of human enteric viruses, protozoa, and indicators of pathogens in the Bagmati River, Nepal. *Pathogens* **2018**, *7*, 38. [CrossRef]
15. Harwood, V.J.; Staley, C.; Badgley, B.D.; Borges, K.; Korajkic, A. Microbial source tracking markers for detection of fecal contamination in environmental waters: Relationships between pathogens and human health outcomes. *FEMS. Microbiol. Rev.* **2014**, *38*, 1–40. [CrossRef]
16. Haramoto, E.; Osada, R. Assessment and application of host-specific *Bacteroidales* genetic markers for microbial source tracking of river water in Japan. *PLoS ONE* **2018**, *13*, e0207727. [CrossRef]
17. Kildare, B.J.; Leutenegger, C.M.; McSwain, B.S.; Bambic, D.G.; Rajal, V.B.; Wuertz, S. 16S rRNA-based assays for quantitative detection of universal, human-, cow-, and dog-specific fecal *Bacteroidales*: A Bayesian approach. *Water Res.* **2007**, *41*, 3701–3715. [CrossRef] [PubMed]
18. Reischer, G.H.; Kasper, D.C.; Steinborn, R.; Mach, R.L.; Farnleitner, A.H. Quantitative PCR method for sensitive detection of ruminant fecal pollution in freshwater and evaluation of this method in alpine karstic regions. *Appl. Environ. Microbiol.* **2006**, *72*, 5610–5614. [CrossRef] [PubMed]
19. Mieszkin, S.; Furet, J.P.; Corthier, G.; Gourmelon, M. Estimation of pig fecal contamination in a river catchment by real-time PCR using two Pig-Specific *Bacteroidales* 16S rRNA genetic markers. *Appl. Environ. Microbiol.* **2009**, *75*, 3045–3054. [CrossRef] [PubMed]
20. Caldwell, J.M.; Raley, M.E.; Levine, J.F. Mitochondrial multiplex real-time PCR as a source tracking method in fecal-contaminated effluents. *Environ. Sci. Technol.* **2007**, *41*, 3277–3283. [CrossRef] [PubMed]
21. Caldwell, J.M.; Levine, J.F. Domestic wastewater influent profiling using mitochondrial real-time PCR for source tracking animal contamination. *J. Microbiol. Methods* **2009**, *77*, 17–22. [CrossRef] [PubMed]
22. Fong, T.T.; Lipp, E.K. Enteric viruses of humans and animals in aquatic environments: Health risks, detection, and potential water quality assessment tools. *Microbiol. Mol. Biol. Rev.* **2005**, *69*, 357–371. [CrossRef] [PubMed]
23. Albinana-Gimenez, N.; Clemente-Casares, P.; Bofill-Mas, S.; Hundesa, A.; Ribas, F.; Girones, R. Distribution of human polyomaviruses, adenoviruses, and hepatitis E virus in the environment and in a drinking-water treatment plant. *Environ. Sci. Technol.* **2006**, *40*, 7416–7422. [CrossRef] [PubMed]
24. Carratalà, A.; Rusinol, M.; Hundesa, A.; Biarnes, M.; Rodriguez-Manzano, J.; Vantarakis, A.; Kern, A.; Suñen, E.; Girones, R.; Bofill-Mas, S. A novel tool for specific detection and quantification of chicken/turkey parvoviruses to trace poultry fecal contamination in the environment. *App. Environ. Microbiol.* **2012**, *78*, 7496–7499. [CrossRef] [PubMed]
25. Hundesa, A.; Maluquer de Motes, C.; Albinana-Gimenez, N.; Rodriguez-Manzano, J.; Bofill-Mas, S.; Suñen, E.; Rosina Girones, R. Development of a qPCR assay for the quantification of porcine adenoviruses as an MST tool for swine fecal contamination in the environment. *J. Virol. Methods* **2009**, *158*, 130–135. [CrossRef] [PubMed]

26. Malla, B.; Ghaju Shrestha, R.; Tandukar, S.; Bhandari, D.; Inoue, D.; Sei, K.; Tanaka, Y.; Sherchand, J.B.; Haramoto, E. Validation of host-specific *Bacteroidales* quantitative PCR assays and their application to microbial source tracking of drinking water sources in the Kathmandu Valley, Nepal. *J. Appl. Microbiol.* **2018**, *125*, 609–619. [CrossRef] [PubMed]
27. World Health Organization (WHO). *Guidelines for Drinking-Water Quality*, 4th ed.; WHO: Geneva, Switzerland, 2011.
28. Constantine, K.; Massoud, M.; Alameddine, I.; El-Fadel, M. The role of water tankers market in water stressed semi-arid urban areas: Implications on water quality and economic burden. *J. Environ. Manag.* **2017**, *188*, 85–94. [CrossRef] [PubMed]
29. Kurokawa, M.; Ono, K.; Nukina, M.; Itoh, M.; Thapa, U.; Rai, S.K. Detection of diarrheagenic viruses from diarrheal fecal samples collected from children in Kathmandu, Nepal. *Nepal Med. Coll. J.* **2004**, *6*, 17–23. [PubMed]
30. Sherchand, J.B.; Schluter, W.W.; Sherchan, J.B.; Tandukar, S.; Dhakwa, J.R.; Choudhary, G.R.; Mahaseth, C. Prevalence of group A genotype human rotavirus among children with diarrhoea in Nepal, 2009–2011. *WHO South-East Asia J. Public Health* **2012**, *1*, 432–440. [CrossRef] [PubMed]
31. Ansari, S.; Sherchand, J.B.; Rijal, B.P.; Parajuli, K.; Mishra, S.K.; Dahal, R.K.; Shrestha, S.; Tandukar, S.; Chaudhary, R.; Kattel, H.P.; et al. Characterization of rotavirus causing acute diarrhoea in children in Kathmandu, Nepal, showing the dominance of serotype G12. *J. Med. Microbiol.* **2013**, *62*, 114–120. [CrossRef] [PubMed]
32. Haramoto, E.; Kitajima, M. Quantification and genotyping of aichi virus 1 in water samples in the Kathmandu Valley, Nepal. *Food Environ. Virol.* **2017**, *9*, 350–353. [CrossRef] [PubMed]
33. Tandukar, S.; Sherchand, J.B.; Karki, S.; Malla, B.; Ghaju Shrestha, R.; Bhandari, D.; Thakali, O.; Haramoto, E. Co-Infection by Waterborne Enteric Viruses in Children with Gastroenteritis in Nepal. *Healthcare* **2019**, *7*, 9. [CrossRef]
34. Shrestha, S.; Haramoto, E.; Shindo, J. Assessing the infection risk of enteropathogens from consumption of raw vegetables washed with contaminated water in Kathmandu Valley, Nepal. *J. Appl. Microbiol.* **2017**, *123*, 1321–1334. [CrossRef]
35. Malla, B.; Ghaju Shrestha, R.; Tandukar, S.; Bhandari, D.; Inoue, D.; Sei, K.; Tanaka, Y.; Sherchand, J.B.; Haramoto, E. Identification of human and animal fecal contamination in drinking water sources in the Kathmandu Valley, Nepal, using host-associated *Bacteroidales* quantitative PCR assays. *Water* **2018**, *10*, 1796. [CrossRef]
36. Malla, B.; Ghaju Shrestha, R.; Tandukar, S.; Sherchand, J.B.; Haramoto, E. Performance evaluation of human-specific viral markers and application of pepper mild mottle virus and crAssphage to environmental water samples as fecal pollution markers in the Kathmandu Valley, Nepal. *Food Environ. Virol.* **2019**. [CrossRef]
37. Uy, D.; Haka, S.; Huya, C.; Srey, M.; Chunhieng, T.; Phoeurng, S.; Nasir, H.M.; Fredricks, D. Comparison of tube-well and dug-well groundwater in the arsenic polluted areas in Cambodia. In *Southeast Asian Water Environment 4*; Fukusi, K., Kurisu, F., Oguma, K., Furumai, H., Fontanos, P., Eds.; International Water Association Publishing: London, UK, 2010.
38. Ferguson, A.S.; Mailloux, B.J.; Ahmed, K.M.; Van Geen, A.; McKay, L.D.; Culligan, P.J. Hand-pumps as reservoirs for microbial contamination of well water. *J. Water Health* **2011**, *9*, 708–717. [CrossRef]
39. Bajracharya, R.; Nakamura, T.; Shakya, B.M.; Kei, N.; Shrestha, S.D.; Tamrakar, N.K. Identification of river water and groundwater interaction at central part of the Kathmandu valley, Nepal using stable isotope tracers. *Int. J. Adv. Sci. Tech. Res.* **2018**, *8*, 29–41. [CrossRef]
40. Ghaju Shrestha, R.; Tanaka, Y.; Malla, B.; Bhandari, D.; Tandukar, S.; Inoue, D.; Sei, K.; Sherchand, J.B.; Haramoto, E. Next-generation sequencing identification of pathogenic bacterial genes and their relationship with fecal indicator bacteria in different water sources in the Kathmandu Valley, Nepal. *Sci. Total Environ* **2017**, *601–602*, 278–284. [CrossRef] [PubMed]
41. Haramoto, E.; Katayama, H.; Asami, M.; Akiba, M. Development of a novel method for simultaneous concentration of viruses and protozoa from a single water sample. *J. Virol. Methods* **2012**, *182*, 62–69. [CrossRef] [PubMed]

42. Haramoto, E.; Kitajima, M.; Hata, A.; Torrey, J.R.; Masago, Y.; Sano, D.; Katayama, H. A review on recent progress in the detection methods and prevalence of human enteric viruses in water. *Water Res.* **2018**, *135*, 168–186. [CrossRef] [PubMed]
43. Kitajima, M.; Hata, A.; Yamashita, T.; Haramoto, E.; Minagawa, H.; Katayama, H. Development of a reverse transcription-quantitative PCR system for detection and genotyping of aichi viruses in clinical and environmental samples. *Appl. Environ. Microbiol.* **2013**, *79*, 3952–3958. [CrossRef] [PubMed]
44. Pal, A.; Sirota, L.; Maudru, T.; Peden, K.; Lewis, A.M. Realtime, quantitative PCR assays for the detection of virus-specific DNA in samples with mixed populations of polyomaviruses. *J. Virol. Methods* **2006**, *135*, 32–42. [CrossRef] [PubMed]
45. Shieh, Y.S.; Wait, D.; Tai, L.; Sobsey, M.D. Methods to remove inhibitors in sewage and other fecal wastes for enterovirus detection by the polymerase chain reaction. *J. Virol. Methods* **1995**, *54*, 51–66. [CrossRef]
46. Katayama, H.; Shimasaki, A.; Ohgaki, S. Development of a virus concentration method and its application to detection of enterovirus and norwalk virus from coastal seawater. *Appl. Environ. Microbiol.* **2002**, *68*, 1033–1039. [CrossRef]
47. Zhang, T.; Breitbart, M.; Lee, W.H.; Run, J.-Q.; Wei, C.L.; Soh, S.W.L.; Hibberd, M.L.; Liu, E.T.; Rohwer, F.; Ruan, Y. RNA viral community in human feces: Prevalence of plant pathogenic viruses. *PLoS Biol.* **2005**, *4*, 108–118. [CrossRef]
48. Haramoto, E.; Kitajima, M.; Kishida, N.; Konno, Y.; Katayama, H.; Asami, M.; Akiba, M. Occurrence of pepper mild mottle virus in drinking water sources in Japan. *Appl. Environ. Microbiol.* **2013**, *79*, 7413–7418. [CrossRef]
49. Jothikumar, N.; Kang, G.; Hill, V.R. Broadly reactive TaqMan assay for real-time RT-PCR detection of rotavirus in clinical and environmental samples. *J. Virol. Methods* **2009**, *155*, 126–131. [CrossRef] [PubMed]
50. Balique, F.; Colson, P.; Barry, A.O.; Nappez, C.; Ferretti, A.; Al Moussawi, K.; Ngounga, T.; Lepidi, H.; Ghigo, E.; Mege, J.-L.; et al. Tobacco mosaic virus in the lungs of mice following intra-tracheal inoculation. *PLoS ONE* **2013**, *8*, e54993. [CrossRef] [PubMed]
51. Heim, A.; Ebnet, C.; Harste, G.; Pring-Åkerblom, P. Rapid and quantitative detection of human adenovirus DNA by real-time PCR. *J. Med. Virol.* **2003**, *70*, 228–239. [CrossRef] [PubMed]
52. Kageyama, T.; Kojima, S.; Shinohara, M.; Uchida, K.; Fukushi, S.; Hoshino, F.B.; Takeda, N.; Katayama, K. Broadly reactive and highly sensitive assay for Norwalk-like viruses based on real-time quantitative reverse transcription-PCR. *J. Clin. Microbiol.* **2003**, *41*, 1548–1557. [CrossRef] [PubMed]
53. Jimenez-Clavero, M.A.; Fernadez, C.; Ortiz, J.A.; Pro, J.; Carbonell, G.; Tarazona, J.V.; Roblas, N.; Ley, V. Teschoviruses as indicators of porcine fecal contamination of surface water. *App. Environ. Microbiol.* **2003**, *69*, 6311–6315. [CrossRef] [PubMed]

pathogens

MDPI

Communication

Weekly Variation of Rotavirus A Concentrations in Sewage and Oysters in Japan, 2014–2016

Erika Ito [1], Jian Pu [2,*], Takayuki Miura [3], Shinobu Kazama [4], Masateru Nishiyama [5], Hiroaki Ito [6], Yoshimitsu Konta [7], Gia Thanh Nguyen [8], Tatsuo Omura [7] and Toru Watanabe [5]

1 The United Graduate School of Agricultural Sciences, Iwate University, Iwate 020-8550, Japan
2 Faculty of Information Networking for Innovation and Design, Toyo University, Tokyo 115-0053, Japan
3 Department of Environmental Health, National Institute of Public Health, Saitama 351-0197, Japan
4 Department of Urban Engineering, The University of Tokyo, Tokyo 113-8656, Japan
5 Department of Food, Life and Environmental Sciences, Yamagata University, Yamagata 997-8555, Japan
6 Center for Water Cycle, Marine Environment and Disaster Management, Kumamoto University, Kumamoto 860-8555, Japan
7 New Industry Creation Hatchery Center, Tohoku University, Miyagi 980-8579, Japan
8 Department of Environmental and Occupational Health, College of Medicine and Pharmacy, Hue University, Hue city 530000, Vietnam
* Correspondence: pu@toyo.jp; Tel.: +81-3-5924-2674

Received: 1 April 2019; Accepted: 23 June 2019; Published: 26 June 2019

Abstract: Concentrations of rotavirus A, in sewage and oysters collected weekly from September 2014 to April 2016 in Japan, were investigated using RT-qPCR; results showed up to 6.5 $\log_{10}$ copies/mL and 4.3 $\log_{10}$ copies/g of digestive tissue (DT) in sewage and oysters, respectively. No correlation was found between rotavirus concentration in sewage and oysters and cases of rotavirus-associated gastroenteritis.

Keywords: rotavirus; oyster; sewage; real-time PCR

1. Introduction

Rotavirus is the major cause of acute gastroenteritis that leads to deaths in infants and young children worldwide. Before vaccines were introduced, rotavirus caused 20–40 deaths annually in the U.S. alone, and mortality was much higher in sub-Saharan Africa and South Asia [1,2]. Moreover, rotavirus was associated with up to 88% of all hospital-associated diarrheal episodes in Japan, before the introduction of vaccines, and led to 2–18 deaths every year [3,4]. While rotavirus can infect all age groups, young groups are mainly affected. Among 4072 rotavirus-associated gastroenteritis cases during the period of 2005–2010 in Japan, approximately 75% were 0- to 2-year-old babies [5]. Various vaccines have been licensed worldwide, including Rotarix, RotaTeq, Rotavac, and Rotasiil [6]. The first two have been commercially available in Japan since November 2011 and July 2012, respectively, for voluntary vaccination. Previous research has shown a decline of rotavirus deaths in 2013, after entering the vaccine era, but mortality in children <5 years remained high globally (197,000–233,000 deaths estimated) [7]. While norovirus has been well recognized to contaminate oysters, causing high levels of gastroenteritis in temperate regions during winter months, rotavirus was also detected in 0.3% to 16.7% of cases with oyster-associated gastroenteritis [8,9]. Although rotavirus has been detected in farmed oysters at rates of 3.3%–44.4% [9–11], information about their level of contamination in the environment and its seasonal variation remains limited. In this study, we performed long-term weekly monitoring of oysters at a cultivation site in Japan, tracking changes in viral loads across different seasons. The incidence of rotavirus in sewage in the same area was also simultaneously monitored, since it is likely to be the main source of rotavirus content in the oysters.

2. Results and Discussion

Data related to rotavirus A contamination in sewage and oyster samples, as well as to gastroenteritis cases, are presented in Figure 1. Among the samples collected between 24 September 2014 and 21 April 2016, the highest rotavirus concentration obtained from sewage and oyster samples was 6.5 $\log_{10}$ copies/mL and 4.3 $\log_{10}$ copies/g of digestive tissue (DT), respectively. Approximately 62.2% (46 of 74 weeks) of sewage and 57.8% (37 of 64 weeks) of oyster samples were positive for rotavirus, which is much higher than the positivity rates reported in previous studies. In Thailand, rotavirus was detected in 27.1% (16 of 59), 9.1% (5 of 55), and 5.4% (5 of 110) of river water, irrigation canal water, and cultured oyster samples, respectively [10]. A wide range of positivity rates for rotavirus has been reported in oysters from different regions. Approximately 3.3% (5 of 150) of farmed oysters in China were found to be contaminated with rotavirus [11], whereas a comparatively higher positivity rate (44.4%, 4 of 9) was found in oysters, related to an outbreak in Southern France [9]. However, we cannot deny the possibility that the positivity rate was influenced by differences in our detection methodologies.

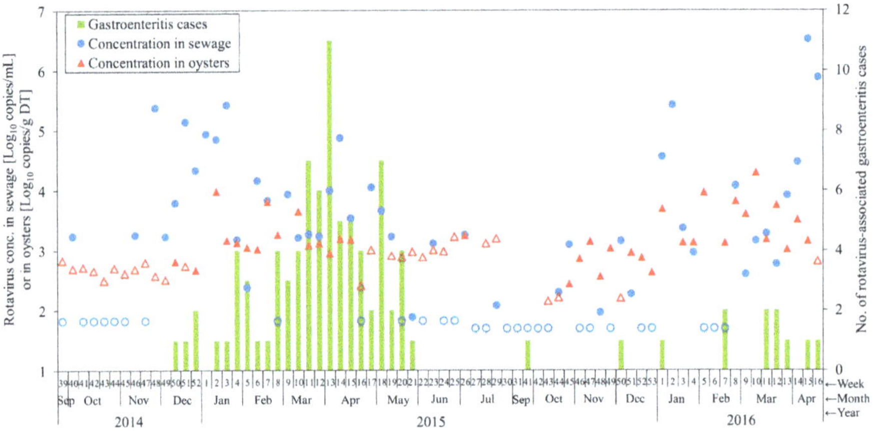

Figure 1. Rotavirus A concentration in sewage and oyster samples together with the number of rotavirus-associated gastroenteritis cases (green columns) in Miyagi, Japan. Empty circles and empty triangles represent half of the detection limit (LOD) in sewage and oysters, respectively, where rotavirus may exist, but below the detection limit. The weeks in which no oyster sample was collected or was tested positive due to low murine norovirus (MNV) recovery rate are considered invalid and left blank; The corresponding number of genomes for quantification cycles (C_q values) of 40 varied across qPCR runs, and the weight of digestive tissue was different in each oyster sample. Thus, LOD for each sewage and oyster sample was different; half of LOD has been shown in the figure for convenience of presentation.

Humans, who consume oysters grown in contaminated water, are at a risk of rotavirus infection. Rotavirus concentrations reached 2.3 $\log_{10}$ PFU/g DT in oysters cultured for 48 h in artificial seawater, containing 10^4 PFU/mL of the rotavirus strain Wa [12]. In Japan, 1 of 286 fecal specimens was found to be positive for rotavirus in 88 oyster-associated gastroenteritis outbreaks [8]. Approximately 16.7% (2 of 12) of patients with shellfish-associated gastroenteritis shed rotaviruses, along with other viruses, such as astrovirus, Aichi virus, and enterovirus [9]. Our cross-correlation analysis found that log transformed norovirus GII concentrations in sewage and oysters was significantly correlated with the number of gastroenteritis cases in the same study area [13]; however, none of the cross-correlation coefficients in this study was statistically significant at the 95% confidence level. There are several possible explanations for this inconsistency. First, the number of rotavirus-associated gastroenteritis cases, reported each week, was small, ranging from 0 to 11, and 56.8% of the weeks (42 of 74) reported no patient with rotavirus-associated disease, according to the Infectious Diseases Weekly Report of Miyagi Prefecture [14]. Second, shedding of rotavirus from domestic animals could cause a high load of rotavirus in seawater and oysters, whereas only those shed by humans could be detected in sewage, since over 99% of animal wastes do not enter municipal sewage system in Japan [15]. On the other hand, infants that receive rotavirus vaccine can shed up to 10^7 copies in one gram of stool [16]; rotavirus vaccine (rotarix)-derived strains were found in six stool samples from pediatric clinics in Japan [17]. Therefore, there is a chance that feces from 5- or 6-month-old vaccinated babies also enter sewage, adding to the complexity of rotaviruses shed from humans. Third, despite the high concentration of rotavirus in seawater, caused by its low removal efficiency by wastewater treatment processes compared to that of norovirus [18,19], different stabilities were observed for different viruses in seawater [20], and different accumulation efficiencies in oysters were observed for different virus strains [21]. This could explain the weak correlations observed in this study. Weekly variation of rotavirus concentrations in sewage and oysters provide new insights into the distribution of rotavirus in wastewater, marine water, and shellfish.

3. Materials and Methods

Municipal sewage (1 L) and oyster (9 in number) samples were collected weekly (73 weeks in total) from Miyagi Prefecture, Japan, between 24 September 2014 and 21 April 2016. Virus particles were concentrated from sewage samples by polyethylene glycol precipitation [22]. Digestive tissue (DT) of each individual oyster was excised, and the virus extracted following a previously described protocol [23]. Approximately 1.5 mL viral supernatant was generated from each oyster. Three supernatants were pooled to form one oyster composite, and 3 oyster composites from each week were used for RNA extraction. Viral RNA was extracted from sewage and oyster samples as described earlier [23]. Complementary DNA (cDNA) was generated via reverse transcription using the iScript Advanced cDNA Synthesis Kit (Bio-Rad, Hercules, CA, USA) and a T100 thermal cycler (Bio-Rad), following the manufacturer's instructions. Rotavirus A was quantified from the cDNAs by quantitative real-time PCR (qPCR) targeting rotavirus on a CFX96 Touch Real-Time PCR Detection System (Bio-Rad), using previously developed primers and probes [24]. Murine norovirus (MNV) was added to samples during the viral extraction step as a whole-process control [22]. Samples with MNV recovery rates higher than 1% were considered valid [25]. Quantification by qPCR was performed in accordance with the minimum information for the publication of real-time quantitative PCR experiments (MIQE) guidelines [26], and samples with quantification cycles (C_q values) below 40 were considered positive for rotavirus.

Lag time (±7 weeks) was studied between log-transformed rotavirus concentrations in sewage and oyster samples (collected weekly) and the number of rotavirus-associated gastroenteritis cases reported weekly by 5 pediatric sentinel clinics in Miyagi Prefecture [15], using cross-correlation analysis [27]. A time-series cross-correlation coefficient of ±7 weeks was calculated to identify correlation between the following events: (1) Occurrence of gastroenteritis cases, (2) shedding of viruses from infected individuals into sewage, and (3) contamination of oysters with viruses. In samples where rotavirus was not detected positively, the incidence of rotavirus was estimated to be half of the limit of detection (LOD) to permit cross-correlation analysis [28,29].

Author Contributions: Conceptualization, T.W. and T.O.; sample process and qPCR analysis of oyster, E.I. and G.T.N.; sample process and qPCR analysis of sewage, S.K. and Y.K.; extraction method of oyster, H.I.; data analysis, J.P. and T.M.; original draft preparation, E.I. and J.P.; editing, J.P., T.M., M.N. and T.W.

Funding: This study was supported by the Japan Science and Technology Agency (JST) through a Core Research for Evolutionary Science and Technology (CREST) program, "Innovation of water monitoring system with rapid, highly precise and exhaustive pathogen detection technologies", and JSPS KAKENHI Grant Number 18H03792.

Acknowledgments: A murine norovirus strain S7-PP3 was kindly provided by Yukinobu Tohya (Nihon University, Japan).

Conflicts of Interest: The authors declare no conflict of interest.

References

1. Desselberger, U. Rotaviruses. *Virus Res.* **2014**, *190*, 75–96. [CrossRef] [PubMed]
2. Estes, M.K.; Greenberg, H.B. Rotaviruses. In *Fields Virology*, 6th ed.; Fields, B.N., Knipe, D.M., Howley, P.M., Eds.; Wolters Kluwer Health/Lippincott Williams Wilkins: Philadelphia, PA, USA, 2013; pp. 1347–1401.
3. Zhou, Y.M.; Li, L.; Kim, B.; Kaneshi, K.; Nishimura, S.; Kuroiwa, T.; Nishimura, T.; Sugita, K.; Ueda, Y.; Nakaya, S.; et al. Rotavirus infection in children in Japan. *Pediatr. Int.* **2000**, *42*, 428–439. [CrossRef] [PubMed]
4. Ministry of Health, Labour and Welfare (MHLW). Rotavirus QAs. Available online: https://www.mhlw.go.jp/bunya/kenkou/kekkaku-kansenshou19/Rotavirus/index.html (accessed on 8 March 2019).
5. National Institute of Infectious Disease (NIID). Report of Rotavirus on 15 May 2013. Available online: https://www.niid.go.jp/niid/ja/diseases/a/echinococcus/392-encyclopedia/3377-rota-intro.html (accessed on 8 March 2019).
6. Desselberger, U. Differences of Rotavirus Vaccine Effectiveness by Country: Likely Causes and Contributing Factors. *Pathogens* **2017**, *6*, 65. [CrossRef] [PubMed]
7. Tate, J.E.; Burton, A.H.; Boschi-Pinto, C.; Parashar, U.D. World Health Organization–Coordinated Global Rotavirus Surveillance Network. Global, Regional, and National Estimates of Rotavirus Mortality in Children <5 Years of Age, 2000–2013. *Clin. Infect. Dis.* **2016**, *62*, S96–S105. [PubMed]
8. Iritani, N.; Kaida, A.; Abe, N.; Kubo, H.; Sekiguchi, J.; Yamamoto, S.P.; Goto, K.; Tanaka, T.; Noda, M. Detection and genetic characterization of human enteric viruses in oyster-associated gastroenteritis outbreaks between 2001 and 2012 in Osaka City, Japan. *J. Med. Virol.* **2014**, *86*, 2019–2025. [CrossRef] [PubMed]
9. Le Guyader, F.S.; Le Saux, J.C.; Ambert-Balay, K.; Krol, J.; Serais, O.; Parnaudeau, S.; Giraudon, H.; Delmas, G.; Pommepuy, M.; Pothier, P.; et al. Aichi virus, norovirus, astrovirus, enterovirus, and rotavirus involved in clinical cases from a French oyster-related gastroenteritis outbreak. *J. Clin. Microbiol.* **2008**, *46*, 4011–4017. [CrossRef]
10. Kittigul, L.; Panjangampatthana, A.; Rupprom, K.; Pombubpa, K. Genetic diversity of rotavirus strains circulating in environmental water and bivalve shellfish in Thailand. *Int. J. Environ. Res. Public Health* **2014**, *11*, 1299–1311. [CrossRef] [PubMed]
11. Kou, X.X.; Wu, Q.P.; Wang, D.P.; Zhang, J.M. Simultaneous detection of norovirus and rotavirus in oysters by multiplex RT–PCR. *Food Control* **2008**, *19*, 722–726. [CrossRef]
12. Araud, E.; DiCaprio, E.; Ma, Y.; Lou, F.; Gao, Y.; Kingsley, D.; Hughes, J.H.; Li, J. Thermal inactivation of enteric viruses and bioaccumulation of enteric foodborne viruses in live oysters (*Crassostrea virginica*). *Appl. Environ. Microbiol.* **2016**, *82*, 2086–2099. [CrossRef]
13. Pu, J.; Miura, T.; Kazama, S.; Konta, Y.; Azraini, N.D.; Ito, E.; Ito, H.; Omura, T.; Watanabe, T. Weekly variations in norovirus genogroup II genotypes in Japanese oysters. *Int. J. Food Microbiol.* **2018**, *284*, 48–55. [CrossRef]
14. Miyagi Prefectural Government (MPG). Infectious Diseases Weekly Report of Miyagi Prefecture. Available online: https://www.pref.miyagi.jp/site/hokans/surveypdf-shuho.html (accessed on 13 March 2018).
15. Ministry of Agriculture, Forestry and Fisheries (MAFF). Survey Results of Animal Waste in Japan. Available online: http://www.maff.go.jp/j/chikusan/kankyo/taisaku/pdf/syori-joukyou.pdf (accessed on 31 May 2019).
16. Hsieh, Y.C.; Wu, F.T.; Hsiung, C.A.; Wu, H.S.; Chang, K.Y.; Huang, Y.C. Comparison of virus shedding after lived attenuated and pentavalent reassortant rotavirus vaccine. *Vaccine* **2014**, *32*, 1199–1204. [CrossRef] [PubMed]
17. Kaneko, M.; Takanashi, S.; Thongprachum, A.; Hanaoka, N.; Fujimoto, T.; Nagasawa, K.; Kimura, H.; Okitsu, S.; Mizuguchi, M.; Ushijima, H. Identification of vaccine-derived rotavirus strains in children with acute gastroenteritis in Japan, 2012–2015. *PLoS ONE* **2017**, *12*, e0184067. [CrossRef] [PubMed]
18. Kitajima, M.; Iker, B.C.; Pepper, I.L.; Gerba, C.P. Relative abundance and treatment reduction of viruses during wastewater treatment processes—Identification of potential viral indicators. *Sci. Total Envron.* **2014**, *488–489*, 290–296. [CrossRef] [PubMed]
19. Miura, T.; Schaeffer, J.; Le Saux, J.C.; Le Mehaute, P.; Le Guyader, F.S. Virus type-specific removal in a full-scale membrane bioreactor treatment process. *Food Environ. Virol.* **2017**, *10*, 176–186. [CrossRef] [PubMed]
20. De Abreu Corrêa, A.; Souza, D.S.; Moresco, V.; Kleemann, C.R.; Garcia, L.A.; Barardi, C.R. Stability of human enteric viruses in seawater samples from mollusc depuration tanks coupled with ultraviolet irradiation. *J. Appl. Microbiol.* **2012**, *113*, 1554–1563. [CrossRef] [PubMed]

21. Le Guyader, F.S.; Atmar, R.L.; Le Pendu, J. Transmission of viruses through shellfish: When specific ligands come into play. *Curr. Opin. Virol.* **2012**, *2*, 103–110. [CrossRef]
22. Kazama, S.; Masago, Y.; Tohma, K.; Souma, N.; Imagawa, T.; Suzuki, A.; Liu, X.; Saito, M.; Oshitani, H.; Omura, T. Temporal dynamics of norovirus determined through monitoring of municipal wastewater by pyrosequencing and virological surveillance of gastroenteritis cases. *Water Res.* **2016**, *92*, 244–253. [CrossRef]
23. Pu, J.; Kazama, S.; Miura, T.; Azraini, N.D.; Konta, Y.; Ito, H.; Ueki, Y.; Cahyaningrum, E.E.; Omura, T.; Watanabe, T. Pyrosequencing analysis of norovirus genogroup II distribution in sewage and oysters: first detection of GII.17 Kawasaki 2014 in oysters. *Food Environ. Virol.* **2016**, *8*, 310–312. [CrossRef]
24. Pang, X.L.; Lee, B.; Boroumand, N.; Leblanc, B.; Preiksaitis, J.K.; Yu Ip, C.C. Increased detection of rotavirus using a real time reverse transcription-polymerase chain reaction (RT-PCR) assay in stool specimens from children with diarrhea. *J. Med. Virol.* **2004**, *72*, 496–501. [CrossRef]
25. ISO 15216-1. Microbiology of the Food Chain-Horizontal Method for Determination of Hepatitis A Virus and Norovirus Using Real-Time RT-PCR- Part 1: Method for Quantification. Available online: https://www.iso.org/standard/65681.html (accessed on 20 March 2018).
26. Bustin, S.A.; Benes, V.; Garson, J.A.; Hellemans, J.; Huggett, J.; Kubista, M.; Mueller, R.; Nolan, T.; Pfaffl, M.W.; Shipley, G.L.; et al. The MIQE guidelines: minimum information for publication of quantitative real-time PCR experiments. *Clin. Chem.* **2009**, *55*, 611–622. [CrossRef]
27. Kazama, S.; Miura, T.; Masago, Y.; Konta, Y.; Tohma, K.; Manaka, T.; Liu, X.; Nakayama, D.; Tanno, T.; Saito, M.; et al. Environmental surveillance of norovirus genogroups I and II for sensitive detection of epidemic variants. *Appl. Environ. Microbiol.* **2017**, *83*, e03406–e03416. [CrossRef] [PubMed]
28. LaFleur, B.; Lee, W.; Billhiemer, D.; Lockhart, C.; Liu, J.; Merchant, N. Statistical methods for assays with limits of detection: Serum bile acid as a differentiator between patients with normal colons, adenomas, and colorectal cancer. *J. Carcinog.* **2011**, *10*, 12. [CrossRef] [PubMed]
29. Whitcomb, B.W.; Schisterman, E.F. Assays with lower detection limits: implications for epidemiological investigations. *Paediatr. Perinat. Epidemiol.* **2009**, *22*, 597–602. [CrossRef] [PubMed]

pathogens

MDPI

Article

Development of a Portable Detection Method for Enteric Viruses from Ambient Air and Its Application to a Wastewater Treatment Plant

Koichi Matsubara and Hiroyuki Katayama *

Department of Urban Engineering, School of Engineering, The University of Tokyo, Tokyo 113-8656, Japan
* Correspondence: katayama@env.t.u-tokyo.ac.jp

Received: 3 August 2019; Accepted: 21 August 2019; Published: 24 August 2019

Abstract: The ambient air from wastewater treatment plants has been considered as a potential source of pathogenic microorganisms to cause an occupational risk for the workers of the plants. Existing detection methods for enteric viruses from the air using a liquid as the collection medium therefore require special care to handle on-site. Knowledge accumulation on airborne virus risks from wastewater has been hindered by a lack of portable and handy collection methods. Enteric viruses are prevalent at high concentrations in wastewater; thus, the surrounding air may also be a potential source of viral transmission. We developed a portable collection and detection method for enteric viruses from ambient air and applied it to an actual wastewater treatment plant in Japan. Materials of the collection medium and eluting methods were optimized for real-time polymerase chain reaction-based virus quantification. The method uses a 4 L/min active air sampler, which is capable of testing 0.7–1.6 m^3 air after 3–7 h sampling with a detection limit of 10^2 copies/m^3 air in the field. Among 16 samples collected at five to seven locations in three sampling trials (November 2007–January 2008), 56% (9/16) samples were positive for norovirus (NV) GII, with the highest concentration of 3.2×10^3 copies/m^3 air observed at the sampling point near a grit chamber. Adenoviruses (4/16), NV GI (6/16), FRNA bacteriophages GIII (3/16), and enteroviruses (3/16) were also detected but at lower concentrations. The virus concentration in the air was associated with that of the wastewater at each process. The results imply that the air from the sewer pipes or treatment process is contaminated by enteric viruses and thus special attention is needed to avoid accidental ingestion of viruses via air.

Keywords: virus; aerosols; pathogenic microorganisms; real-time PCR

---

1. Introduction

Wastewater treatment plants are considered as potential sources of pathogenic bioaerosols [1]. Several studies have demonstrated that high amounts of microorganisms are present not only in the wastewater but also in bioaerosols generated from wastewater treatment processes [2–4]. Bioaerosols are suspected to have adverse health effects on the neighboring residents of wastewater treatment processes [2] or wastewater treatment plant (WWTP) workers [5]; however, there are limited studies about the detection of enteric viruses from bioaerosols [2,6]. These studies used cell culture assays and detected enterovirus (EV) and reovirus which did not reflect the actual occurrence of viruses because most viruses are practically difficult to propagate in cell lines. Moreover, the research field for bioaerosol monitoring has predominately focused on the detection of fungal spores and bacteria, where the analysis of samples depends on total-count or culture techniques.

Enteric viruses are shed in the feces of infected patients; thus, they are frequently detected at high concentrations in wastewater samples [7]. They are transmitted mainly through the fecal–oral route via contaminated food and water, but some epidemiological reports have shown that enteric viruses, especially noroviruses (NVs), can cause outbreaks through aerosols released from vomit [8,9]. Quantitative polymerase chain reaction (qPCR) has been widely used to detect enteric viruses in wastewater because some enteric viruses such as human NVs cannot routinely be propagated in cell lines [10]. Furthermore, the PCR assays have the advantages of specificity, sensitivity, and rapidity in the detection; hence, this can be a reliable method for detecting viruses in bioaerosols.

A previous study detected noroviruses from the air using dust filter (PTFE filter with the pore size 1 μm), while the method was not optimized for virus detection and the detection rate was low (only one in four field samples) [4]. Another recent study detected rotavirus and adenovirus (AdV) quantitatively with a liquid collector and cascade sampler using PCR [11]. However, knowledge is limited partly due to the complicated sampling method. The lack of a portable collection method hampers knowledge accumulation on airborne virus risks from wastewater. The reliable existing method uses liquid for collection [12], but this is not convenient for sampling as it requires a regular power supply (AC 100–200 V), which is not always available in the field or specific locations of WWTPs.

Also, the liquid medium requires special care to be handled on-site to avoid contamination. Collection media for air sampling is vulnerable to contamination since viruses that originate from wastewater are abundant in the environment in WWTPs. Operation at an unevenly leveled location or transportation from the field to the laboratory can also cause the liquid to spill from its container. There was also an attempt to use membranes for sampling in previous literature [6]. However, it was not optimized for detecting viruses and for PCR detection processes. Therefore, it is important to develop and test a handy, battery-driven sampling method using a membrane optimized for qPCR.

The objective of this study was to develop a mobile sampling device and sampling procedure for the detection of enteric viruses in bioaerosols by PCR-based assay, and to apply the method at an actual WWTP. In this course, we developed a novel mobile sampling method and verified it via field sampling at an actual WWTP.

2. Materials and Methods

2.1. Development of Collection Method and Laboratory Evaluation

A mixed cellulose membrane (HA 0.45 μm, Millipore) was used as collection media with glycine buffer (pH 9.5) to elute the viruses as previously tested among various membrane materials and pore sizes [13]. The membrane was placed in a sterilized 47 mm monitor holder. The HA membrane was proofed to be effective in collecting enteric viruses in the water sample and in eluting the viruses in alkaline solution [14]. The developed method was evaluated by comparing it with an existing liquid collection method (Figure 1). For the liquid collector, we used an SKC Biosampler in which the air was in contact with the liquid circulating inside the container as the standard collection device for viruses among various liquid collectors [12]. The SKC Biosampler was operated with a vacuum pump at a flow rate of 12.5 L/min. In this experiment, two pumps were prepared separately; one for bubbling the viruses and another for sampling such that the airflow rate for bubbling was the same between the newly developed method and the SKC Biosampler. F-specific RNA coliphage Qbeta [15], Poliovirus (LSc-2ab Sabin strain), and murine NV (S7-PP3 strain, isolated in Japan) were used to test the recovery media. The coliphages were propagated in bacterial host *Eschelichia coli* (*E. coli*) K-12 F+ (A/λ) in LB broth solution, followed by filtration with the membrane (pore size 0.45 μm). Poliovirus and murine NV (MNV) were propagated in RAW264.7 cells as previously described [16] to obtain 4.8×10^9–3.4×10^{11} plaque-forming units per mL. The titer of the phage and virus stock solution was determined by plaque assay using a double agar overlay method. Then the virus stock solutions (0.1–10 mL) were inoculated into 1 L of sterilized phosphate buffer solution. A 100 mL portion of the inoculated solution was aerated in a 250 mL gas washing bottle by a vacuum pump at a flow rate of 4–12 L/min to generate the virus-containing aerosols. A portable sampling mini-pump (Shibata) and low-volume air sampler (AirCheck HV30) were used for aspiration. The generated aerosols were transported by silicon tubes directly to the collection apparatuses.

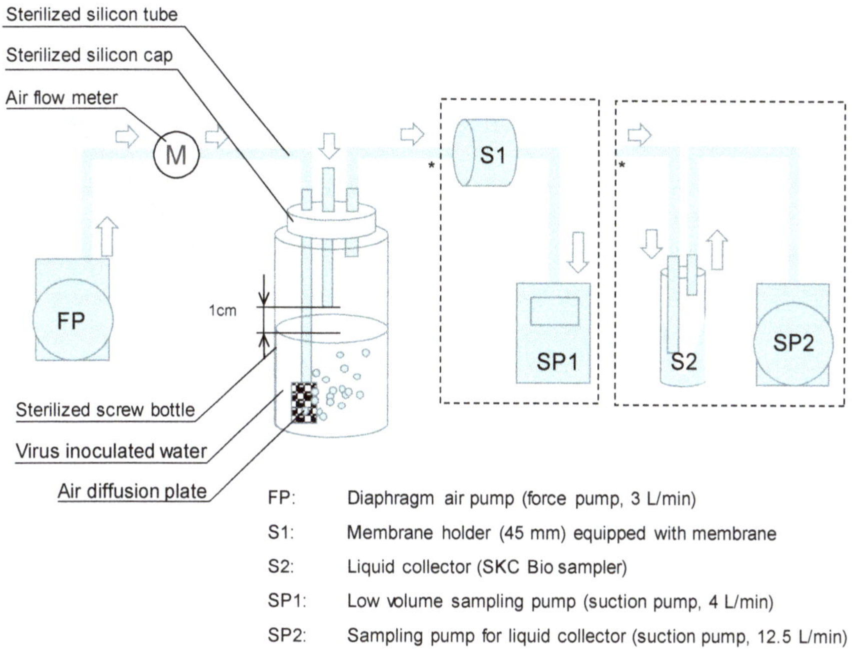

Figure 1. Experimental settings for comparison with the existing sampling method.

2.2. *qPCR Assay*

A 140 µL of the eluate was used for the RNA extraction process by Qa IAamp viral RNA mini kit (Qiagen, Japan) to obtain a final volume of 60 µL. Then the samples were subjected to a reverse transcription step using the High Capacity cDNA RT kit with RNase (Applied Biosystems) following the manufacturer's protocol. Five microliter portions of cDNAs were quantified by real-time quantitative PCR using the ABI PRISM 7500 sequence detection system (Applied Biosystems). The sequence of primers and probes [14,17–19] and thermal conditions of real-time PCR are shown in Table S1.

2.3. *Sampling at the Wastewater Treatment Plant*

Air and water samples were collected from a wastewater treatment plant located in Japan with a capacity of 450,000 m^3/day, adopting the conventional activated sludge (AS) treatment. The treatment process consists of a grit chamber, AS chamber, final settlement chamber, and chlorine contact chamber. The wastewater was mostly made up of domestic sewage since the catchment area of the plant was residential. The plant was located in a residential area and thus all treatment facilities were in the buildings to prevent the odor from getting outside. The AS chamber was covered and equipped with an exhaust duct. The air in the exhaust duct from the AS chamber and the grit chamber was treated with a wet type air scrubber (mist separator) followed by a biological deodorization chamber and an activated carbon deodorization chamber. The treated air was exhausted from an exhaust tower at a public park.

Figure 2 represents the sampling points at each treatment process. Air at the AS chamber (A) was taken by silicon tube from an inspection hole on top of the chamber cover. The sampling point of the air was approximately 80 cm above the liquid surface of the AS. Air from the exhaust duct (B) and treated air (D) was also taken by silicon tube from an odor inspection hole of the duct. The drain sample from the wet type air scrubber (C) was taken in liquid form. Points F and G were right above the inflow screen of the grit chamber. Point F was at the floor level on the grating while G was sampled at 1.2 m above the ground using a tripod (Figure S1). Ambient air in the AS building (E) and the grit chamber building (H) were also sampled at a 1.2 m height from the ground. Access to points F and G was controlled only for the workers, while points E and H were in the middle of the factory, accessible to visitor tours. There was no other source of droplets or aerosols of wastewater than those at the sampling points. All the sampling points were inside the building where the outside wind did not affect the sampling procedure. The air temperature inside the building was not an extreme condition; the temperature at point A was measured to be 25.0 °C (November 2007), 17.6 °C (December 2007), and 16.5 °C (January 2008). The ambient air temperature was measured at some sampling points for reference purposes, recording 24.2 °C at site E in November 2007 and 13.5 °C at Site F in January 2008.

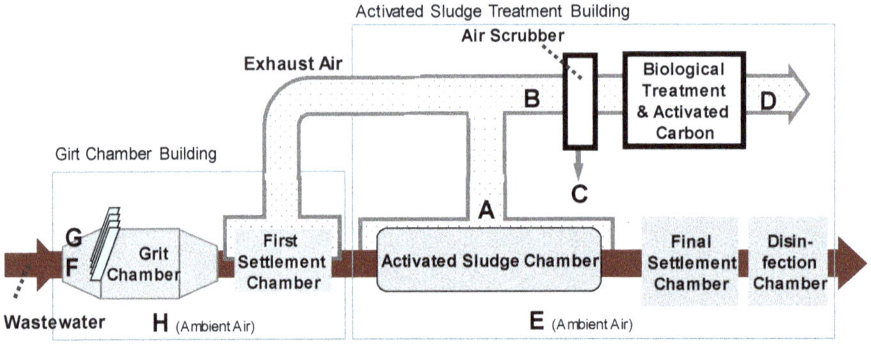

Figure 2. Schematic of wastewater treatment plant treatment flow and sampling points. A; Activated sludge chamber, B; Exhaust air duct, C; Drainage of mist separator, D; Treated air, E; Ambient air at activated sludge building, F; Grit chamber (Floor Level + 0 m near wastewater inflow screen), G; Grit chamber (Floor Level + 1.2 m near wastewater inflow screen), H; Ambient air at grit chamber building.

3. Results and Discussion

3.1. Evaluation of the Developed Sampling Method

From the three trials for both the developed HA vortex method and the liquid collector, the captured virus amounts showed similar collection capacities; there were no significant differences for captured viruses per m^3 of air between the methods (Table 1). The recovery ratio was not obtained because the dispersion ratio of viruses from the bubbled water was unknown. MNV tended to be recovered more than EV and bacteriophages (Qβ). The difference in recovery by the virus species may be due to the mechanisms by which viruses were transported from the liquid to air. The tendency for the transportation of the virus from the aqueous phase to air is unclear on the laboratory scale. The previous study showed that the hydrophobicity of the enteric viruses was different among species [20]. The hydrophobic particles are more likely to be aerosolized (transported to air–water interfaces). Results from the field also support that the difference in transportation capacity from seawater to air among various taxa of bacteria was due to different levels of hydrophobicity [21].

Table 1. Comparison of recovery ratio between developed method (HA vortex) and liquid collector.

Trial	Method	Concentration of Viruses in the Bubbled Water Sample (Copies/mL Water)						Dissipated Water Volume (mL)	Air Volume (m^3)	Captured Viruses per Air (Copies/m3 Air)		
		Qβ		PV		MNV				Qβ	PV	MNV
		Before	After	Before	After	Before	After					
1	HA vortex	2.9×10^8	2.7×10^8	2.5×10^6	1.6×10^6	7.7×10^5	1.5×10^6	2.06	121	1.3×10^2	2.5×10^0	1.9×10^3
	Liquid Collector							4.33	375	6.7×10^1	2.1×10^1	3.2×10^2
2	HA vortex	3.1×10^8	3.3×10^8	2.4×10^6	1.6×10^6	1.4×10^6	7.2×10^5	2.01	121	5.4×10^1	9.5×10^0	3.9×10^2
	Liquid Collector							3.52	375	4.9×10^1	2.3×10^1	1.6×10^2
3	HA vortex	3.3×10^8	2.9×10^8	2.4×10^6	2.1×10^6	1.4×10^6	7.2×10^5	2.02	121	3.5×10^1	1.2×10^1	3.5×10^2
	Liquid Collector							4.37	375	5.5×10^1	2.1×10^1	2.8×10^2

Note: Qβ; Bacteriophage, PV; Poliovirus, MNV; Murine norovirus, Before/After; Virus concentration of the virus-inoculated water sample before/after bubbling (virus generation) manipulation.

3.2. Application to Wastewater Treatment Plant

Table 2 shows the results of virus detection from the air and the AS or raw sewage at all sampling points and periods. AdV, NV GI, and NV GII were detected in all water and sludge samples. The detection rate of viruses in the air at sampling points A, B, and F was 89% (8/9) (Table S2). NV GII showed the highest concentration among the viruses tested, with the highest concentration observed at the grit chamber (F, 6.0×10^2 copies/m^3 in geometric mean, n = 3), followed by the AS chamber (A, 2.4×10^2 copies/m^3, n = 2).

Table 2. Virus concentrations in sewage and activated sludge.

Trial	Sample Water Type	Virus Concentration, Copies/mL Water				
		AdV	NV GI	NV GII	FG3	EV
Nov-07	Activated Sludge	1.1×10^2	4.1×10^1	1.0×10^3	1.7×10^3	1.2×10^2
Dec-07	Activated Sludge	9.7×10^1	7.4×10^2	9.4×10^3	4.0×10^2	2.5×10^2
	Drain from Mist Separator	5.2×10^{-1}	1.7×10^0	1.5×10^2	3.8×10^0	+
Jan-08	Activated Sludge	1.6×10^3	5.5×10^2	1.4×10^4	4.0×10^2	1.6×10^3
Jan-08	Raw Sewage	4.4×10^3	4.1×10^2	2.5×10^4	1.3×10^3	2.5×10^3
Jan-08	Drain from Mist Separator	ND	1.1×10^1	9.4×10^2	1.1×10^1	7.8×10^0

Note: +; Detected but not quantified. NV GI; Norovirus genogroup I, NV GII; Norovirus genogroup II, AdV; Adenovirus (all serotypes), FG3; F+-specific RNA coliphage serotype 3.

Figure 3 shows the quantified virus concentration at each site. The exhaust air shows high NV GII concentration (1.0×10^2 copies/m^3, n = 2) before air treatment (B), but NV was not detected after air treatment (D). EV was detected only once from the post-treatment air, though the level was below the quantification limit. Exhaust air treatment effectively reduced the virus in the air, which was also supported by the fact that the viruses were detected from the drain of the air scrubber (Table 2). Viruses (NV GI, NV GII, EV and FG3) were observed in the ambient air in the grit chamber building (H, 6.3×10^2 copies/m^3). Sampling location H was in the aisle, where there was no machinery or wastewater surface within 2–3 m. The detection of viruses in distant locations such as E and H suggests that the viruses may be aerosolized and dispersed in the building.

Detected locations and sampling periods of NV GI and GII were consistent. For instance, the result of NV GII was always positive if that of NV GI was positive (6/6). Furthermore, NV GI showed the highest concentration at points F and G in December 2017, when the NV GII concentration was the highest. NV GI was quantified in only two samples. On the other hand, AdV was observed in the sample in which NV GII was not detected. In this sample, false-negative results may have been obtained for NV GII because of the problem in the sampling method (see Section 3.4, a comparison with a liquid sampler, for discussion).

High virus concentration at the grit chamber building implies that risk is relatively higher at the place that is in contact with raw sewage, as compared to the location of the treatment process in the AS chamber. Virus aerosols may be supplied from raw water pipes because of the pumping at the upstream of the pipeline, or aerosolized at the mechanical stress at the mechanical screen, which is the location where regular monitoring and maintenance is necessary to precwent garbage from clogging the screen. The risk for WWTP workers is normally controlled because the maintenance personnel usually wear masks and other protective equipment. However, those risk control measures have not been evaluated considering the possible ingestion of enteric viruses from the air. Although our results do not give a comprehensive risk evaluation, they at least show that the protection measures at the grit chamber or near the raw sewage inflow should be prioritized to avoid the unintended ingestion of enteric viruses from sewage.

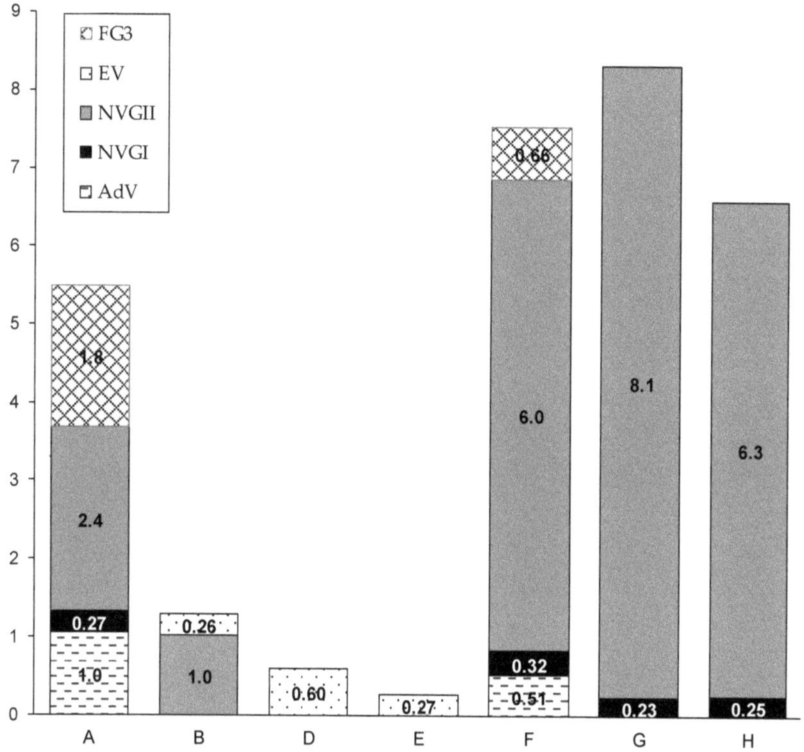

Figure 3. Virus detection at the wastewater treatment plant. Note: All data are shown in geometric means of several sampling trials (see Figure S2 for detailed data). NV GI; Norovirus genogroup I, NV GII; Norovirus genogroup II, AdV; Adenovirus (all serotypes), FG3; F+-specific RNA coliphage serotype 3.

3.3. Comparison of Virus Concentration in the Water and Air

The virus concentration in the AS was compared with that in the air at sampling points near the wastewater or the AS (A, B, and F) (Figure 4). There was a moderate to strong correlation between the log-transformed virus concentration in the liquid phase of the AS and the virus concentration in the air ($r = 0.74$, Pearson's correlation test $p < 0.001$).

The overall correlation between the air and the liquid phase implies that the virus concentration in the air was quantified properly. Given that all conditions in the AS are controlled, the virus concentration in the air should be well correlated with the liquid phase. It is true that the correlation is not always consistent; the NV GII concentration in the AS was higher in January than in December, while the concentration in the air (sampling point A) was lower in January. This result may imply that the virus concentration in the AS was not the only factor to be transmitted through the air. For instance, the strength of aeration in the AS chamber may increase the rate of the virus droplet or aerosol generation. Further study is needed to substantiate the accuracy of the virus detection method in controlled settings in a laboratory experiment.

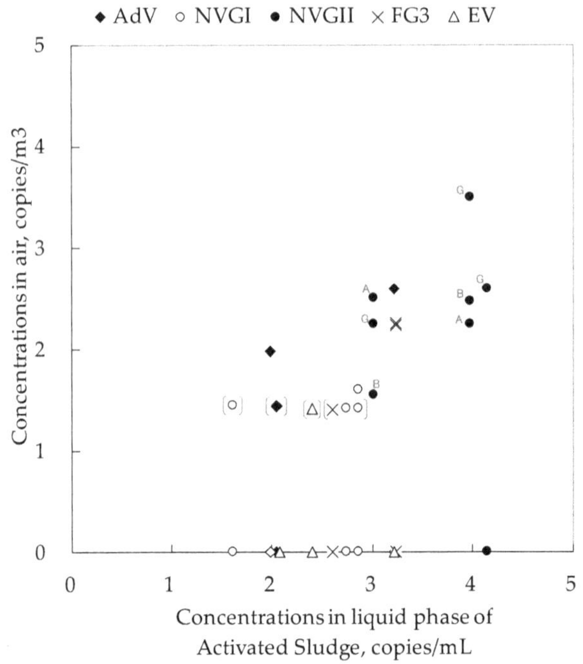

Figure 4. Correlations between virus concentration in activated sludge (liquid phase) and in the air. Notes: The plots on the horizontal axis represent samples not detected. [ ]: Under the quantification limit (detected but not quantified; the plot gives the detection limit value). NV GI; Norovirus genogroup I, NV GII; Norovirus genogroup II, AdV; Adenovirus (all serotypes), FG3; F+-specific RNA coliphage serotype 3.

3.4. Comparison with Liquid-Based Sampler

The results in the two sampling periods (December 2007 and January 2008) from the air above the AS chamber (at point A) were compared (Table 3). There are cases in which either our method or the liquid collector was negative while the other was positive.

Table 3. Comparison between the developed method (HA vortex) and liquid collector.

Sampling Period	Site A (Copies/mL)					
	HA Vortex			Liquid Collector		
	NV GI	NV GII	FGIII	NV GI	NV GII	FGIII
December-07	+	1.8×10^2	-	-	+	-
January-08	-		-	-	1.2×10^3	3.1×10^3

Note: + Detected but not quntified.

There was a case in which our method was not as good as the liquid collector. For instance, NV GII was detected by both the HA vortex method and the liquid collector in December 2007, while it was detected only by the liquid collector in January 2008. It was observed at the laboratory that, two to three hours after the sampling, only the membrane sample at site A in January 2008 was wet. The surface of the membrane was obviously wet and its color was changed (slightly transparent). The reason for the wet membrane condition may be due to increased water droplets from the AS caused by the operating conditions, or the aeration intensity of the AS chamber may have been high. The wetness obstructed the membrane's pores and possibly the air shortcut in the apparatus. Higher humidity in the air may also have affected the collection ratio because humidity moisturizes the membrane surface and may change the electrostatic condition of the membrane. From this result, it should be noted that our collection method may not be stable if the environmental conditions change. Alternatively, the diameter of droplets and aerosols may affect the collection efficiency because the pore size of the membrane was much larger than the viruses. In this sense, a negative result for the virus cannot guarantee the absence of viruses in the air. On the other hand, only our HA vortex method detected NV GI and a higher concentration of NV GII in December 2007.

The results show that our HA vortex method is comparable to the existing liquid collector concerning the detection of viruses in the air. In addition, the HA vortex method was capable of sampling the air for a longer period than the liquid collector; the liquid collector is normally capable of sampling for only 30 min for fear the liquid will evaporate. Although further study is needed to overcome the false-negative case, it can be easily avoided since it was visibly identifiable by the wet condition of the membrane after sampling. Also, the wet condition of the membrane only happened in a case where samples were taken from an AS chamber with actively aerated water, which thus produced many water droplets. This condition is not likely to arise in ambient air sampling. Therefore, considering sampling ease, our method is superior to the existing method because the sampling media is a solid membrane that can easily be handled and transported.

4. Conclusions

We developed a mobile virus collection method for sampling enteric viruses from the air (HA vortex method), which was optimized for detection by PCR. The method was confirmed to have a similar virus collection ability to that of the existing collection method using liquid media during the laboratory test. Further study is, however, required to improve the collection method since it showed a false-negative result in a field sample when the membrane was wet. The failed case was visibly identifiable due to the membrane surface conditions. To assure the reproducibility of the results, careful checking the membrane condition after the sampling is necessary when the method is applied to a highly humid sampling location. Despite this weakness, the portable detection method we presented has a high potential for the detection of viruses both in the laboratory and the field. The collection media is solid, light, and handy and the method avoids on-site manipulation, which may pose a risk of contamination to the samples. The method has an advantage for WWTP sampling because of the virus-abundant nature of its environment. The developed handy and portable method will encourage the study of enteric viruses from the air, which is an understudied research topic.

The developed method was applied to a WWTP in Japan and successfully detected enteric viruses in the air which pose an occupational risk for the wastewater treatment plant workers. As the air scrubber removes the viruses dispersed from the WWTP, the risk to neighbors can be controlled by conventional odor control measures. Among all the treatment processes, NV GII was detected in the highest frequency and concentration at the grit chamber. The research suggests that the air near raw sewage has a higher risk of dispersing viruses than the air generated by treatment processes such as AS. It is recommended that appropriate protective measures be taken against the unintended ingestion of enteric viruses from the air, especially near raw sewage.

Supplementary Materials: The following are available online at http://www.mdpi.com/2076-0817/8/3/131/s1, Table S1: Sequences of primers and probes for real-time PCR, Table S2: Detection of viruses from the wastewater treatment plant (detail), Figure S1: Photo of sampling (at the wastewater treatment plant).

Author Contributions: Conceptualization, K.M. and H.K.; methodology, K.M.; validation, H.K.; formal analysis, K.M.; investigation, K.M.; resources, H.K.; data curation, K.M.; writing—original draft preparation, K.M.; writing—review and editing, H.K.; visualization, K.M.; supervision, H.K.; project administration, H.K.; funding acquisition, H.K.

Funding: This research received no external funding.

Conflicts of Interest: The authors declare no conflict of interest.

References

1. Adams, A.P.; Spendlove, J.C. Coliform aerosols emitted by sewage treatment plants. *Science* **1970**, *169*, 1218–1220. [CrossRef] [PubMed]
2. Fannin, K.F.; Vana, S.C.; Jakubowski, W. Effect of an activated sludge wastewater treatment plant on ambient air densities of aerosols containing bacteria and viruses. *Appl. Environ. Microbiol.* **1985**, *49*, 1191–1196. [PubMed]
3. Sawyer, L.A.; Murphy, J.J.; Kaplan, J.E.; Pinsky, P.F.; Chacon, D.; Walmsley, S.; Schonberger, L.B.; Phillips, A.; Forward, K.; Goldman, C.; et al. 25 to 30nm virus particle associated with a hospital outbreak of acute gastroenteritis with evidence for airborne transmission. *Am. J. Epidemiol.* **1988**, *127*, 1261–1271. [CrossRef] [PubMed]
4. Uhrbrand, K.; Schultz, A.C.; Madsen, A.M. Exposure to Airborne Noroviruses and Other Bioaerosol Components at a Wastewater Treatment Plant in Denmark. *Food Environ. Virol.* **2011**, *3*, 130–137. [CrossRef]
5. Orsini, M.; Laurenti, P.; Boninti, F.; Arzani, D.; Lanni, A.; Romano-Spica, V. A molecular typing approach for evaluating bioaerosol exposure in wastewater treatment plant workers. *Water Res.* **2002**, *36*, 1375–1378. [CrossRef]
6. Wallis, C.; Melnick, J.L.; Rao, V.C.; Sox, T.E. Method for detecting viruses in aerosols. *Appl. Environ. Microbiol.* **1985**, *50*, 1181–1186. [PubMed]
7. Haramoto, E.; Katayama, H.; Oguma, K.; Yamashita, H.; Tajima, A.; Nakajima, H.; Ohgaki, S. Seasonal profiles of human noroviruses and indicator bacteria in a wastewater treatment plant in Tokyo, Japan. *Water Sci. Technol.* **2006**, *54*, 301–308. [CrossRef] [PubMed]
8. Marks, P.J.; Vipond, I.B.; Regan, F.M.; Wedgwood, K.; Fey, R.E.; Caul, E.O. A school outbreak of Norwalk-like virus: Evidence for airborne transmission. *Epidemiol. Infect.* **2003**, *131*, 727–736. [CrossRef] [PubMed]
9. Marks, P.J.; Vipond, I.B.; Carlisle, D.; Deakin, D.; Fey, R.E.; Caul, E.O. Evidence for airborne transmission of Norwalk-like virus (NLV) in a hotel restaurant. *Epidemiol. Infect.* **2000**, *124*, 481–487. [CrossRef] [PubMed]
10. Bartnicki, E.; Cunha, J.B.; Kolawole, A.O.; Wobus, C.E. Recent advances in understanding noroviruses. *F1000Research* **2017**, *6*, 79. [CrossRef] [PubMed]
11. Brisebois, E.; Veillette, M.; Dion-Dupont, V.; Lavoie, J.; Corbeil, J.; Culley, A.; Duchaine, C. Human viral pathogens are pervasive in wastewater treatment center aerosols. *J. Environ. Sci.* **2018**, *67*, 45–53. [CrossRef] [PubMed]
12. Hogan, C.J.; Kettleson, E.M.; Lee, M.H.; Ramaswami, B.; Angenent, L.T.; Biswas, P. Sampling methodologies and dosage assessment techniques for submicrometre and ultrafine virus aerosol particles. *J. Appl. Microbiol.* **2005**, *99*, 1422–1434. [CrossRef] [PubMed]
13. Matsubara, K.; Haramoto, E.; Katayama, H.; Ohgaki, S. Detection method for enteric viruses in bioaerosol from sewage treatment plant with quantitative real-time PCR. In Proceedings of the 14th International Symposium on Health-Related Water Microbiology, Tokyo, Japan, 9–15 September 2007; pp. 321–322.
14. Katayama, H.; Shimasaki, A.; Ohgaki, S. Development of a virus concentration method and its application to detection of enterovirus and norwalk virus from coastal seawater. *Appl. Environ. Microbiol.* **2002**, *68*, 1033–1039. [CrossRef] [PubMed]
15. Limsawat, S.; Ohgaki, S. Fate of liberated viral RNA in wastewater determined by PCR. *Appl. Environ. Microbiol.* **1997**, *63*, 2932–2933. [PubMed]
16. Kitajima, M.; Oka, T.; Tohya, Y.; Katayama, H.; Takeda, N.; Katayama, K. Development of a broadly reactive nested reverse transcription-PCR assay to detect murine noroviruses, and investigation of the prevalence of murine noroviruses in laboratory mice in Japan. *Microbiol. Immunol.* **2009**, *53*, 531–534. [CrossRef] [PubMed]
17. Heim, A.; Ebnet, C.; Harste, G.; Pring-Åkerblom, P. Rapid and quantitative detection of human adenovirus DNA by real-time PCR. *J. Med. Virol.* **2003**, *70*, 228–239. [CrossRef] [PubMed]
18. Ogorzaly, L.; Gantzer, C. Development of real-time RT-PCR methods for specific detection of F-specific RNA bacteriophage genogroups: Application to urban raw wastewater. *J. Virol. Methods* **2006**, *138*, 131–139. [CrossRef] [PubMed]
19. Kageyama, T.; Kojima, S.; Shinohara, M.; Uchida, K.; Fukushi, S.; Hoshino, F.B.; Takeda, N.; Katayama, K. Broadly reactive and highly sensitive assay for Norwalk-like viruses based on real-time quantitative reverse transcription-PCR. *J. Clin. Microbiol.* **2003**, *41*, 1548–1557. [CrossRef] [PubMed]

20. Imai, T.; Sano, D.; Miura, T.; Okabe, S.; Wada, K.; Masago, Y.; Omura, T. Adsorption characteristics of an enteric virus-binding protein to norovirus, rotavirus and poliovirus. *BMC Biotechnol.* **2011**, *11*, 123. [CrossRef] [PubMed]
21. Michaud, J.M.; Thompson, L.R.; Kaul, D.; Espinoza, J.L.; Richter, R.A.; Xu, Z.Z.; Lee, C.; Pham, K.M.; Beall, C.M.; Malfatti, F.; et al. Taxon-specific aerosolization of bacteria and viruses in an experimental ocean-atmosphere mesocosm. *Nat. Commun.* **2018**, *9*, 2017. [CrossRef] [PubMed]

pathogens

MDPI

Article

Fecal Source Tracking in A Wastewater Treatment and Reclamation System Using Multiple Waterborne Gastroenteritis Viruses

Zheng Ji [1,2], Xiaochang C. Wang [2], Limei Xu [2], Chongmiao Zhang [2], Cheng Rong [2], Andri Taruna Rachmadi [3], Mohan Amarasiri [4], Satoshi Okabe [5], Naoyuki Funamizu [6] and Daisuke Sano [3,4,*]

1 National Demonstration Center for Experimental Geography Education, School of Geography and Tourism, Shaanxi Normal University, Xi'an 710119, China; jizheng@snnu.edu.cn
2 Key Laboratory of Northwest Water Resource, Ecology and Environment, Ministry of Education, Shaanxi Key Laboratory of Environmental Engineering, Xi'an University of Architecture and Technology, Xi'an 710055, China; xcwang@xauat.edu.cn (X.C.W.); Xulimei@sdau.edu.cn (L.X.); cmzhang@xauat.edu.cn (C.Z.); chenrong@xauat.edu.cn (C.R.)
3 Department of Frontier Science for Advanced Environment, Graduate School of Environmental Studies, Tohoku University, Aoba 6-6-06, Aramaki, Aoba-ku, Sendai, Miyagi 980-8579, Japan; andri.rachmadi@kaust.edu.sa
4 Department of Civil and Environmental Engineering, Graduate School of Engineering, Tohoku University, Aoba 6-6-06, Aramaki, Aoba-ku, Sendai, Miyagi 980-8579, Japan; mohan.amarasiri.b5@tohoku.ac.jp
5 Graduate School of Engineering, Hokkaido University, North 13, West 8, Kita-ku, Sapporo, Hokkaido 011-Mizumoto-cho 27-1, Muroran, Hokkaido 060-8628, Japan; sokabe@eng.hokudai.ac.jp
6 Department, Muroran Institute of Technology, Mizumoto-cho 27-1, Muroran, Hokkaido 050-8585, Japan; n_funamizu@mmm.muroran-it.ac.jp
* Correspondence: daisuke.sano.e1@tohoku.ac.jp; Tel.: +81-11-795-7481

Received: 22 August 2019; Accepted: 27 September 2019; Published: 30 September 2019

Abstract: Gastroenteritis viruses in wastewater reclamation systems can pose a major threat to public health. In this study, multiple gastroenteritis viruses were detected from wastewater to estimate the viral contamination sources in a wastewater treatment and reclamation system installed in a suburb of Xi'an city, China. Reverse transcription plus nested or semi-nested PCR, followed by sequencing and phylogenetic analysis, were used for detection and genotyping of noroviruses and rotaviruses. As a result, 91.7% (22/24) of raw sewage samples, 70.8% (17/24) of the wastewater samples treated by anaerobic/anoxic/oxic (A^2O) process and 62.5% (15/24) of lake water samples were positive for at least one of target gastroenteritis viruses while all samples collected from membrane bioreactor effluent after free chlorine disinfection were negative. Sequence analyses of the PCR products revealed that epidemiologically minor strains of norovirus GI (GI/14) and GII (GII/13) were frequently detected in the system. Considering virus concentration in the disinfected MBR effluent which is used as the source of lake water is below the detection limit, these results indicate that artificial lake may be contaminated from sources other than the wastewater reclamation system, which may include aerosols, and there is a possible norovirus infection risk by exposure through reclaimed water usage and by onshore winds transporting aerosols containing norovirus.

Keywords: waterborne gastroenteritis viruses; fecal source tracking; wastewater reclamation; viral contamination

1. Introduction

Wastewater treatment and reclamation systems using membrane technologies such as membrane bioreactor (MBR) are becoming increasingly employed in mitigating the shortage of clean water sources [1,2]. However, usage of reclaimed wastewater may increase the exposure risk of humans to pathogenic microorganisms, if the wastewater treatment system is not capable of effectively removing these microorganisms [3].

Indicator microorganisms are available to assess and guarantee the microbiological quality of water, because the presence of such indicator microorganisms points to the possible existence of similar pathogens and represents a failure in the treatment system which affects the final effluent [4,5]. Fecal indicator bacteria (FIB) (total coliforms, fecal coliforms, *Escherichia coli*, fecal streptococci and spores of sulphite-reducing *clostridia*) have been used to assess the water quality and treatment performance for decades [5]. However, FIB could not identify the sources of the contamination and there are many complexities related to the extra-enteric ecology of FIBs including environmental persistence and particle association [6,7]. It is unclear how to estimate the contribution of different sources of feces when sources are mixed, which would further hinder the water quality management and health risk evaluation.

As an alternative, specific microbial source-tracking (MST) markers have been suggested as suitable indicators for evaluating the contamination and treatment performance. crAssphage is one of the suggested human specific contamination markers and found to have geographical and temporal differences [8,9]. *Bacteroidales* and *Lachnospiraceae* which contain host-specific microorganisms are also suggested as alternative indicators [10]. Some studies have suggested waterborne gastroenteritis viruses as MST markers due to their prevalence in host feces and stringent host specificity [11–14] which provides information on pathogen status that is not provided by indicator bacteria and bacteriophages [6].

Even though usage of gastroenteritis viruses as MST markers in evaluating the fecal contamination has been documented, studies in evaluating the suitability of viral indicators to evaluate treatment unit performance are scarce. Especially, in systems like MBR which use size separation as one major virus removal mechanism, microbes with larger diameter sizes (>1 μm), including bacteria (FIB included) and protozoa, can be effectively removed with microfiltration while viral pathogens which are smaller than bacterial pathogens (< 100 nm) could easily pass through the MBR facilities if they are not attached to larger particles, and are much more environmentally resistant than the indicator bacteria [15–18]. It is further evinced by the absence of correlations between FIB and enteric viruses in MBR effluents [19,20]. Therefore, it is necessary to identify waterborne gastroenteritis viruses circulating in membrane-based wastewater reclamation systems which can be used as indicators to evaluate the treatment unit performance to ensure that reclaimed wastewater is microbiologically safe and not posing infectious risks.

In this study, phylogenetic analysis of multiple waterborne gastroenteritis viruses was applied to estimate contamination sources in a wastewater treatment and reclamation system with a hybrid process of anaerobic/anoxic/oxic (A^2O) combined with a membrane bioreactor (MBR). Noroviruses and rotaviruses were selected because they were of great significance in disease transmission [21]. The extent of the viral pollution in the system was evaluated by the frequency of positive samples for viral genes from the wastewater samples. The genetic diversity of these viruses was determined by nucleotide sequencing and phylogenetic analysis in order to identify prevalent genotypes and their persistence, which were the underlying evidence for estimating the contamination sources of these gastroenteritis viruses. To the best of our knowledge, a comprehensive study of this kind, by the inclusion of human viruses in wastewater, has rarely before been performed in northwestern China.

2. Results

2.1. Occurrence of Viral Genes in Wastewater Samples

We analyzed the quantity of human norovirus GI, GII and rotavirus and their removal in a wastewater treatment plant utilized in a University Campus. Wastewater influent contained septic tank effluents, kitchen wastewater and greywater. Wastewater was treated using fine screen, A^2O treatment and MBR. Effluent wastewater was discharged in to a recreational lake.

Concentration of complex environmental samples might also simultaneously concentrate the PCR inhibitory substances, thus resulting in interference in virus detection. To increase sensitivity, the nested/semi-nested PCR was employed. The results of inhibition test indicated that PCR inhibitors possibly existing in wastewater did not affect the virus detection from the collected samples (data not shown). The occurrences of viruses in samples collected from different sites were summarized in Table 1. High level of fecal contamination in the study area was revealed by the high percentages of positive samples for norovirus and rotavirus. After analyzing 96 wastewater samples, norovirus GI and GII were found in 52% (50/96) and rotavirus in 32% (31/96) of samples (Table 1).

Table 1. Occurrences of waterborne gastroenteritis viruses in wastewater samples.

Virus	Sampling Locations % (Positive/Total Samples)				Total Detection Rate for Each Virus (%)
	Mixed Raw Sewage	A^2O Effluent	MBR Effluent after Disinfection	Lake Water	
HuNoV GI	67 (16/24)	45 (11/24)	0 (0/24)	38 (9/24)	38 (36/96)
HuNoV GII	79 (19/24)	50 (12/24)	0 (0/24)	33 (8/24)	41 (39/96)
HRVs	75 (18/24)	29 (7/24)	0 (0/24)	25 (6/24)	32 (31/96)
Total Detection Rate for Each Sampling Site (%)	92 (22/24)	71 (17/24)	0 (0/24)	63 (15/24)	56 (54/96)

The number of viruses detected in wastewater samples from different sites was variable. Only one virus was detected in 16% (15/96) of samples, including 5 raw sewage samples, 4 A^2O effluent samples and 6 lake water samples. More than one virus type was found in 29% (28/96) of samples, including 16 raw sewage samples, 7 A^2O effluent samples, and 5 lake water samples. These indicate that different families of gastroenteritis viruses are co-circulating in the study area. For mixed raw sewage collected after the fine screen, 22 samples (92%) were positive for viruses; norovirus GI/GII was found in 83% (20/24) and rotavirus in 75% (18/24). Gastroenteritis viruses in raw sewage must have originated from black water from toilet flushing and grey water from washing, which are potentially contaminated by feces or vomit from infected humans. For the A^2O effluent samples, 17 samples representing 71% (17/24) were positive; norovirus were found in 71% (17/24) and rotavirus in 29% (7/24). For lake water, 14 (58%) samples were positive for viruses. Norovirus was found in 54.2% (13/24) while rotavirus was found in 25.0% (6/24).

2.2. *Phylogenetic Analysis of Norovirus*

The norovirus sequences detected in wastewater samples were distributed between the two genogroups. 72% (36/50) of the sequences were similar to GI while 78% (39/50) belonged to GII, whereas 50% (25/50) of them were positive for both GI and GII. Figures 1 and 2 illustrate the result of phylogenetic analysis for capsid region in norovirus genes obtained from wastewater samples. Multiple genotypes of norovirus (GI.3, GI.4, GI.6, GII.3, GII.4 (Den Haag), GII.6 and GII.13) circulating in the study area between human populations and wastewater were detected. The high similarity in identities between norovirus genes detected from multiple samples collected from different sampling sites in this area might suggest that the samples might be contaminated by human noroviruses from the same original source—the residents in the study area.

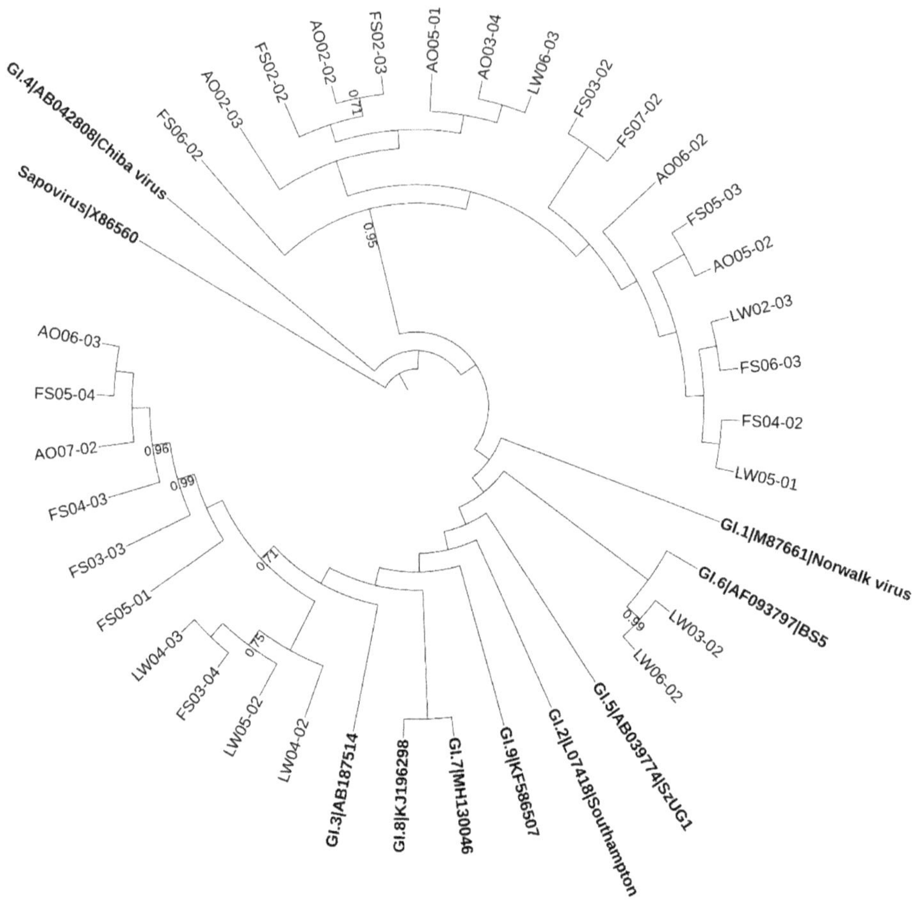

Figure 1. The phylogenetic tree based on partial sequences of the capsid gene of norovirus GI. The tree was constructed by the maximum-likelihood method with 1000 bootstrap replicates using MEGA7 and depicted using iTOL4. The obtained sequences were expressed as the abbreviation of sampling site + month + time. Numbers at each branch indicate bootstrap values for the clusters supported by that branch (>0.7). Numbers at each branch indicate bootstrap values for the clusters supported by that branch. Sapovirus was used as an out group. Reference sequences are shown in bold face.

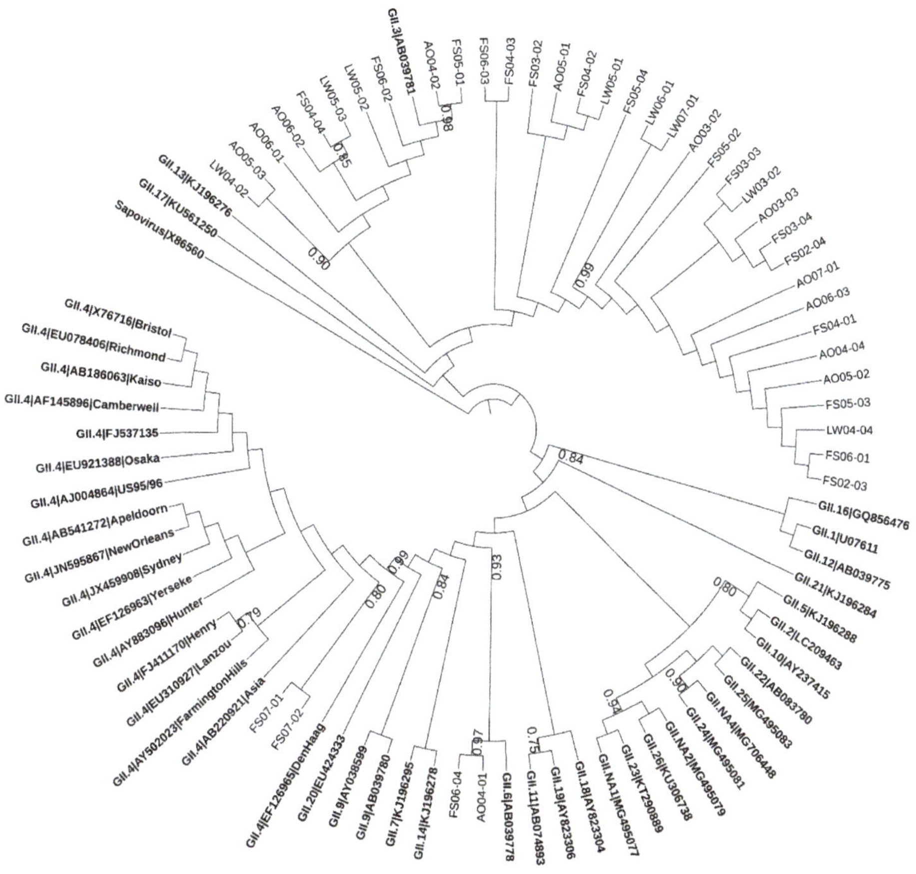

Figure 2. The phylogenetic tree based on partial sequences of the capsid gene of norovirus GII. The tree was constructed by the maximum-likelihood method with 1000 bootstrap replicates using MEGA7 and depicted using iTOL4. The obtained sequences were expressed as the abbreviation of sampling site + month + time. Numbers at each branch indicate bootstrap values for the clusters supported by that branch (>0.7). Numbers at each branch indicate bootstrap values for the clusters supported by that branch. Sapovirus was used as an out group. Reference sequences are shown in bold face.

2.3. Molecular Detection and Characterization of Rotavirus

Group A rotavirus has been shown to be the most prevalent rotavirus in children and adults over the world [22,23]. Therefore, these viruses are considered of great epidemiological importance. Human rotaviruses (HRVs) were characterized with genotype-specific primers for VP7 (G genotype). The phylogenetic analysis was performed for the PCR products derived from wastewater samples (Figure 3), which indicated that all clones were highly homologous to human rotavirus isolates. The most frequent G type detected was type G9, followed by G2 and G3.

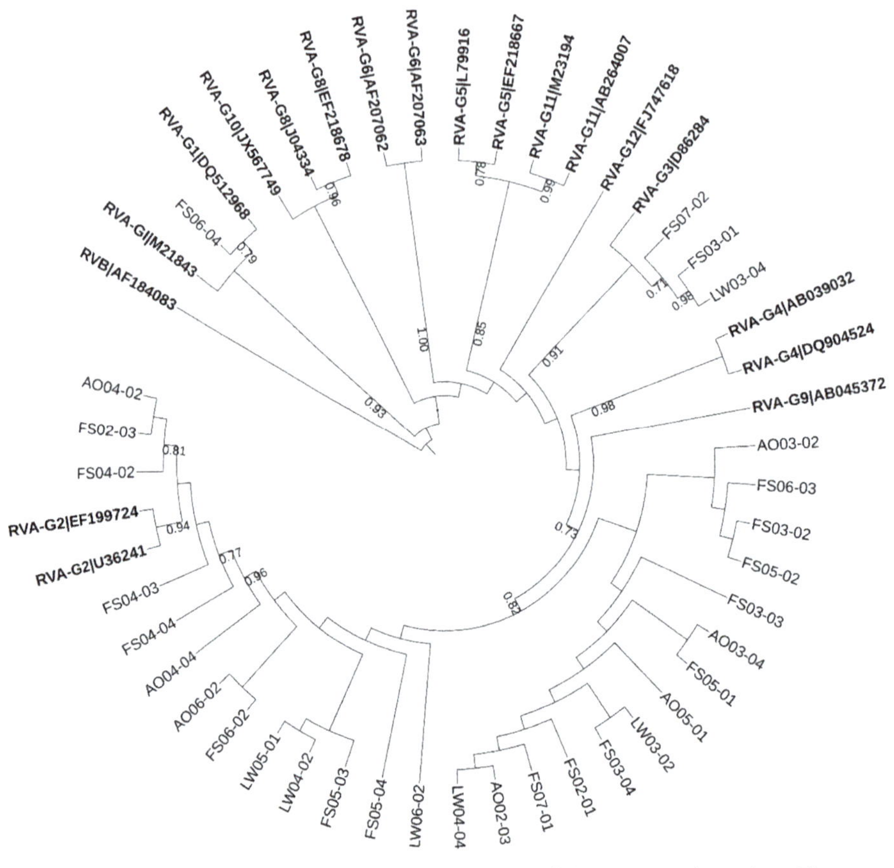

Figure 3. The phylogenetic tree based on partial sequences of the VP7 gene of rotavirus. The tree was constructed by the maximum-likelihood method with 1000 bootstrap replicates using MEGA7 and depicted using iTOL4. The obtained sequences were expressed as the abbreviation of sampling site + month + time. Numbers at each branch indicate bootstrap values for the clusters supported by that branch (>0.7). Numbers at each branch indicate bootstrap values for the clusters supported by that branch. Human Rotavirus B (RVB) was used as an out group. Reference sequences are shown in bold face.

3. Discussion

We confirmed the presence of human noroviruses (GI and GII) and rotaviruses in the influent wastewater, fine screen effluent, A^2O treatment effluent, and the lake water receiving the wastewater effluents. The lower virus detection rate observed after the A^2O treatment process compared to raw sewage may be owing to the attachment to wastewater solids and the presence of antiviral components in the activated sludge [24–27]. Gastroenteritis viruses were not detectable in the samples of MBR effluent after free chlorine disinfection. MBR combined with chlorine treatment may have significantly contributed to the reduction of virus particles, or at least the MBR with chlorine treatment may decrease the virus quantity to a very low extent which was below the detection limit [28]. However, 54% of the lake water samples were positive for viruses, implying that the MBR effluent disinfected with free chlorine may not be the source of virus contamination in the lake water.

The results of phylogenetic analysis revealed that the artificial lake was contaminated by multiple human viruses. In this case, sewage pipe leakage and overflows are not likely to cause such contamination due to the adequately designed capacity and the proper maintenance of the water distribution system. Secondary contamination of lake water may occur from unidentified nonpoint sources. As the lakes are open water bodies in the local water system, they were vulnerable to contamination generating from natural processes (such as surface runoff, water air transfer and wild animals) or human activities [29,30]. As non-point sources of gastroenteritis viruses, rain water inflow and aerosol blowing into the lakes may be considered as possible reasons. Furthermore, it would be of particular concern because the microbial aerosols containing viral particles could be formed during water reclamation, and exposure to reclaimed water can pose a potential health risk [31]. On the other hand, onshore winds around 4 m/s can contain $5.3 \pm 1.2 \times 10^4$ m^{-3} of viruses [32]. These results underscore the possible impact of viral exposure by reclaimed water consumption, and by being exposed to winds containing aerosols and suggests that the control of non-point viral sources, and storage and safe use of reclaimed water should be the focus of wide attention.

The sequence diversity of human noroviruses, especially for the capsid region, from environmental samples has been reported in several studies [33–35]. The isolation of both GI and GII strains in this study would indicate the co-existence of extensive recessive infections for both genogroups which may not be included and documented in previous epidemiological surveys. However, results similar to our present study have been obtained in some environmental studies [36,37]. Thus it might indicate a distinct genogroup prevalent bias between clinical samples and environmental samples [38,39]. It has been demonstrated that the viral loads of GI in fecal samples was reported less than one percent of that of GII and GI is generally more resistant to wastewater treatment and disinfection than GII [38,39], suggesting the differences in environmental occurrence and persistence of GI and GII strains [40]. Although there was no documentation about the viral infection in the studying area, the report of Xi'an Center for Disease Control and Prevention showed that HuNoV GII was more prevalent than HuNoV GI in clinical samples (data not shown). However, human norovirus strains detected in wastewater may reflect more accurate actual circulation among population rather than clinical survey, because wastewater receive viruses shed from patients with both symptomatic and asymptomatic infections. Thus, the findings indicate the possibility that norovirus GI strains might be more widely spread among humans than previously thought. Other explanations such as seasonal or geographic variation in viral RNA levels could not be excluded either.

Number of rotavirus A genotypes (G1, G2, G3, and G9) were detected during the sampling period and G9 was predominant. Previous surveys confirmed the circulation of multiple rotavirus A genotypes in the same area in the same year [22] even though the predominant rotavirus genotype varied in different geographical regions [41–43]. The phylogenetic analysis of rotavirus also suggests that the viruses detected in this study might originate from infant, children or healthy carriers, and thus their contamination sources or transport routes could be different from those of fecal indicators usually originating from adults.

It has been recognized that enteric viruses are more stable than indicator bacteria in water and sewage, constituting not only a potential hazard but also a good tracer for fecal pollution source tracking [14,44,45]. Wastewater treatment plants (WWTPs) have played an important role in microbiological reduction, minimizing the risks associated with pathogen circulation into the environment [3,18]. However, little is known about the comparative persistence or survival of source-specific markers and strains, and the available data for markers ranging from *E. coli* to *Bacteroidales* and phage markers indicate strongly that survival is not proportional [46]. The general trend is that the dominance of environmental strains differ from strains in the host. Due to the inherent difficulty in finding a correlation between environmental contamination and cases of infection, microbiological monitoring of the environment might be more helpful for source tracking and water safety control rather than risk assessment [47,48]. In addition, limited waterborne viral outbreaks usually occurred at distance from the original source of contamination. This study provides novel evidence of the prevalence and genetic diversity of waterborne gastroenteritis viruses and the potential of human noroviruses for microbial source tracking due to its host-specificity and higher sensitivity of (semi-)nested PCR (detection about 10^0 copies/reaction) [49,50]. Attention should be paid to the emerging health threat due to the different predominant types of the targeting viruses observed in the study.

Furthermore, although direct sequencing analysis with well-purified PCR amplicons could be useful for providing information on viral identification in wastewater [37], the potential that the results may have a bias in interpreting the genetic diversity of the viral types might not be neglected. This might be resulted from the inhibition effect as the recovery rate of water concentration [3] and the affinity selection of PCR reaction might be type and strain different for viruses [51]. This more comprehensive analysis of the relative abundance and occurrence of viruses in wastewaters may allow for the development of more conservative viral tracers and complementary indicators to further ensure the microbial safety of wastewater reclamation systems.

4. Materials and Methods

4.1. Sample Collection

To investigate waterborne gastroenteritis viral pollution, four kinds of wastewater samples were collected four times per month for a 6-month sampling period (from Feb. to Jul., 2012, the total sample number is 96) in a wastewater treatment and reclamation system in Xi'an Si-yuan University. The university is located in the south-eastern suburb of Xi'an in Northwest China. WWTP is a hybrid of anaerobic/anoxic/oxic (A^2O) combined with a membrane bioreactor (MBR) (As shown in Figure 4) [52,53]. The influent is a mixture of black water from toilet flushing, grey water from miscellaneous uses, and kitchen wastewater from the university canteens. The reclaimed water is supplied to the lakes in the campus which have both the functions of landscaping and storage reservoirs where the water is further supplied to buildings for toilet flushing and/or to the green belt for gardening and irrigation. All samples were collected on clear weather days, stored in sterilized plastic bottles on ice, and delivered to the laboratory within several hours after collection.

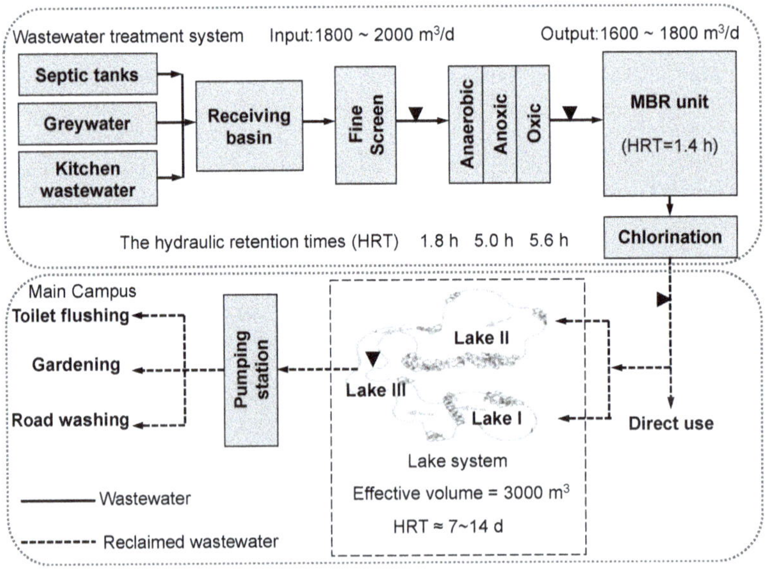

Figure 4. Sampling locations in the local wastewater treatment and reclamation system. Four types of wastewater samples were mixed raw sewage samples collected after fine screen (FS), the effluent of A^2O treatment tank (AO), MBR effluent after disinfection (MBR) and lake water (LW).

4.2. Recovery of Viral Particles and Nucleic Acid Extraction

Since the density of waterborne gastroenteritis viruses is presumed to be very low in water, an efficient viral concentration method is required [49]. It is important to recognize that there is no single method yet by which it is possible to recover all enteric viruses with high efficiency from diverse types of water samples [49]. On the basis of the properties of urban sewage and viral particles, the methods of aqueous polymer two-phase separation (polyethylene glycol precipitation, PEG precipitation) and/or virus adsorption elution (VIRADEL) using electronegative membrane filters (mixed cellulose ester) were applied to concentrate viruses from different types of wastewater samples in the study [49]. For high turbidity (>100 NTU) samples such as raw sewage collected after the fine screen and the effluent of A^2O treatment tank, 250 mL of each was concentrated by PEG precipitation method [54,55]. For low turbidity (<100 NTU) samples such as the effluent of MBR and the lake water, 2 L of each was concentrated by VIRADEL method [56] followed by PEG precipitation. Viral concentrates were resuspended in 1 mL distilled deionized water (DDW) and immediately processed for nucleic acid extraction or stored at −80 °C until use.

Viral RNA was extracted from sample concentrates with QIAamp® Viral RNA Mini Kit (Qiagen, Hilden, Germany), following the manufacturer's instructions. Complementary DNA (cDNA) was synthesized from 10 µL out of 60 µL of the extracted RNA with DNase treatment and subsequent reverse transcription (RT) reaction using PrimeScript® RT reagent Kit with gDNA Eraser (Takara, Dalian, China) according to the protocol described by the manufacturer. The synthesized cDNA was stored at −80 °C for further analysis.

4.3. Molecular Detection and Characterization of Enteric Viruses

The detection and characterization of waterborne gastroenteritis viruses were performed with a combination of several molecular techniques which allowed both sensitive and precise identification of predominant human pathogenic viruses occurring in urban sewers. The capsid encoding region with higher host-specificity was chosen for nested or semi-nested PCR detection of HuNoVs and HRVs (Table 2). The molecular characterization of HuNoVs and HRVs was performed by sequencing and phylogenetic analysis of the second round of PCR amplicons. For the first PCR round, 2 µL of cDNA was added to a reaction mixture consisting of 0.25 µL of Ex Taq (Takara, Dalian, China), 2.5 µL of 10× Ex Taq Buffer, 2 µL of deoxynucleoside triphosphate (dNTP) mixture, and 400 nM of each PCR primer, and all mixed with DDW to obtain a total volume of 25 µL. For the second PCR round, the same concentration of reagents was used with 2 µL of 1000-fold dilution of the first PCR product added to the PCR tube. Primer sequences and positions, and cycling conditions for detection and characterization of each viral group are shown in Table 1. Positive and negative controls (clinical samples for each virus type and RNA/DNA-free water) were included in all PCR runs. PCR products were analyzed by gel electrophoresis on a 1.5% (wt/vol) strength agarose gel, stained with GelRed™ Nucleic Acid gel stain (Biotium, Fremont, CA, USA), and visualized by UV illumination. When no amplification products were observed, two-fold and four-fold dilutions of the identical wastewater sample were prepared and applied to the nested/semi-nested RT-PCR for checking the presence of PCR inhibition. As the reference, 1 mL of DDW added with 1 µL virus suspension was used.

Table 2. Primers and amplification conditions used for detection and molecular characterization of waterborne gastroenteritis viruses.

Virus	Target Gene	PCR Round	Primer	Sequence (5′-3′) [a]	Reference
Rotavirus	VP7(G)	1st	RoA [b]	CTTTAAAAGAGAGAATTTCCGTCTG	[57,58]
		1st	RoB [b]	TGATGATCCCATTGATATCC	
		2nd	RoC [b]	TGTATGGTATTGAATATACCAC	
		2nd	RoD [b]	ACTGATCCTGTTGGCCAWCC	
Norovirus GI	ORF1–ORF2 junction	1st	COG1F [c]	CGYTGGATGCGNTTYCATGA	[34,59]
		1st	G1-SKR [c]	CCAACCCARCCATTRTACA	
		2nd	G1-SKF [c]	CTGCCCGAATTYGTAAATGA	
		2nd	G1-SKR [c]	CCAACCCARCCATTRTACA	
Norovirus GII	ORF1–ORF2 junction	1st	COG2F [d]	CARGARBCNATGTTYAGRTGGATGAG	[34,59]
		1st	G2-SKR [e]	CCRCCNGCATRHCCRTTRTACAT	
		2nd	G2-SKF [e]	CNTGGGAGGGCGATCGCAA	
		2nd	G2-SKR [e]	CCRCCNGCATRHCCRTTRTACAT	

[a] Mixed bases in degenerate primers are as follows: K = G/T; M = A/C; R = A/G; S = G/C; W = A/T; Y = C/T; B = G/T/C; H = A/T/C; N = A/T/G/C; [b] Corresponding nucleotide position of HRV (K02033) of the 5′ end; [c] Corresponding nucleotide position of HuNoV (M87661) of the 5′ end; [d] Corresponding nucleotide position of HuNoV (AF145896) of the 5′ end; [e] Corresponding nucleotide position of HuNoV (X86557) of the 5′ end. Rotavirus, 1st PCR: 94 °C for 3 min; 35 cycles of 94 °C for 30 s, 37 °C for 30 s, and 72 °C for 1 min; and 72 °C for 5 min; 2nd PCR: 94 °C for 3 min; 35 cycles of 94 °C for 30 s, 37 °C for 30 s, and 72 °C for 30 s; and 72 °C for 5 min. Norovirus, 94 °C for 5 min; 40 cycles of 94 °C for 30 s, 50 °C for 30 s, and 72 °C for 30 s; and 72 °C for 10 min.

4.4. Nucleotide Sequencing and Phylogenetic Analysis

PCR products obtained from the second round of amplification for each virus group were excised from the gel and purified immediately. The purified nucleotides were sent to Sangon Biotech (Shanghai, China) Co., Ltd for sequence determination. After checking the sequence chromatograms with Chromas software (version 2.31) for errors, the final sequences were obtained. Homology searches were conducted using the GenBank server of the National Centre for Biotechnology Information (NCBI) and the Basic Local Alignment Search Tool (BLAST) algorithm and calicivirus typing tool (https://norovirus.phiresearchlab.org/). Phylogenetic relationships were generated using maximum likelihood method using MEGA 7 by Kimura 2-parameter model with nucleotide substitution rates following a gamma-distribution. One thousand bootstrap replications were performed to evaluate the robustness of each node [60–62]. Interactive Tree Of Life (iTOL) v4 was used to develop the phylogenetic trees [63].

4.5. Nucleotide Sequence Accession Numbers

The nucleotide sequences corresponding to fragments of rotaviruses and noroviruses have been deposited in the GenBank database under accession No. KF854668 to KF854698 and KF854593 to KF854667, respectively.

5. Conclusions

In conclusion, this study describes novel findings on the prevalence and genetic diversity of human gastroenteritis viruses in water in China. It confirmed that human fecal contamination is widespread and also that viral tools are applicable as fecal indicators and tracers in all geographical areas studied. Continuous viral contamination monitoring is useful for preventing waterborne disease outbreaks and for understanding the impact caused by human activities and the use of reclaimed wastewater.

Furthermore, this study highlights the importance of further environmental studies toward a better understanding of the circulation of gastroenteritis viruses in aquatic environments and human populations. In other words, circulation of gastroenteritis viruses between contaminated environmental water and human populations is a key issue in understanding their epidemiology and health risks for humans. Further studies are needed to define the relationship between the level of gastroenteritis viruses contamination detected by PCR in reclaimed wastewater and the potential effect and health risk of these wastewater after consumption.

Author Contributions: Conceptualization, X.C.W., N.F., and D.S.; methodology, Z.J., L.X., C.Z., and D.S.; software, A.T.R. and M.A.; validation, C.R., S.O., and D.S.; formal analysis, Z.J. and M.A.; investigation, Z.J., L.X., A.T.R., and M.A.; resources, X.C.W.; data curation, Z.J., L.X., C.Z., and D.S.; writing—original draft preparation, Z.J.; writing—review and editing, M.A. and D.S.; visualization, M.A.; supervision, X.C.W.; project administration, X.C.W.; funding acquisition, X.C.W.

Funding: This work was supported by the Strategic China-Japan Joint Research Program on "S&T for Environmental Conservation and Construction of a Society with Less Environmental Burden" (NSFC Grant No. 51021140002), the National Key Research and Development Program of China (No. 2017YFE0127300), the National Natural Science Foundation of China (No. 51578441), the Natural Science Foundation of Shaanxi Province (No. 2017JQ5074), the Research Plan of Science and Technology Department of Xi'an city (2017071CG/RC034(SXSF002)), the Fundamental Research Funds for the Central University (No. GK201603075, GK201601009 and GK201802108) and the Youth Innovation Team Project in the Geography and Tourism College of Shaanxi Normal University. The APC was funded by Japan Science and Technology Agency.

Conflicts of Interest: The authors declare no conflict of interest.

References

1. Angelakis, A.N.; Asano, T.; Bahri, A.; Jimenez, B.E.; Tchobanoglous, G. Water reuse: From ancient to modern times and the future. *Front. Environ. Sci.* **2018**, *6*, 26. [CrossRef]
2. Zhang, C.-M.; Wang, X.-C. Distribution of enteric pathogens in wastewater secondary effluent and safety analysis for urban water reuse. *Hum. Ecol. Risk Assess.* **2014**, *20*, 797–806. [CrossRef]
3. Sano, D.; Amarasiri, M.; Hata, A.; Watanabe, T.; Katayama, H. Risk management of viral infectious diseases in wastewater reclamation and reuse: Review. *Environ. Int.* **2016**, *91*, 220–229. [CrossRef] [PubMed]
4. Bofill-Mas, S.; Rusiñol, M.; Fernandez-Cassi, X.; Carratalà, A.; Hundesa, A.; Girones, R. Quantification of human and animal viruses to differentiate the origin of the fecal contamination present in environmental samples. *BioMed Res. Int.* **2013**, *2013*, 192089. [CrossRef]
5. García-Aljaro, C.; Blanch, A.R.; Campos, C.; Jofre, J.; Lucena, F. Pathogens, faecal indicators and human-specific microbial source-tracking markers in sewage. *J. Appl. Microbiol.* **2019**, *126*, 701–717. [CrossRef] [PubMed]
6. Field, K.G.; Samadpour, M. Fecal source tracking, the indicator paradigm, and managing water quality. *Water Res.* **2007**, *41*, 3517–3538. [CrossRef]
7. O'Mullan, G.D.; Dueker, M.E.; Juhl, A.R. Challenges to managing microbial fecal pollution in coastal environments: Extra-enteric ecology and microbial exchange among water, sediment, and air. *Curr. Pollut. Rep.* **2017**, *3*, 1–16.
8. García-Aljaro, C.; Ballesté, E.; Muniesa, M.; Jofre, J. Determination of crAssphage in water samples and applicability for tracking human faecal pollution. *Microb. Biotechnol.* **2017**, *10*, 1775–1780. [CrossRef] [PubMed]
9. Edwards, R.A.; Vega, A.A.; Norman, H.M.; Ohaeri, M.; Levi, K.; Dinsdale, E.A.; Cinek, O.; Aziz, R.K.; McNair, K.; Barr, J.J.; et al. Global phylogeography and ancient evolution of the widespread human gut virus crAssphage. *Nat. Microbiol.* **2019**, 527796. [CrossRef]
10. Mclellan, S.L.; Eren, A.M. Discovering new indicators of fecal pollution. *Trends Microbiol.* **2014**, *22*, 697–706. [CrossRef]
11. Wong, K.; Fong, T.-T.; Bibby, K.; Molina, M. Application of enteric viruses for fecal pollution source tracking in environmental waters. *Environ. Int.* **2012**, *45*, 151–164. [CrossRef] [PubMed]
12. Harwood, V.J.; Boehm, A.B.; Sassoubre, L.M.; Vijayavel, K.; Stewart, J.R.; Fong, T.-T.; Caprais, M.-P.; Converse, R.R.; Diston, D.; Ebdon, J.; et al. Performance of viruses and bacteriophages for fecal source determination in a multi-laboratory, comparative study. *Water Res.* **2013**, *47*, 6929–6943. [CrossRef] [PubMed]
13. Rachmadi, A.T.; Torrey, J.R.; Kitajima, M. Human polyomavirus: Advantages and limitations as a human-specific viral marker in aquatic environments. *Water Res.* **2016**, *105*, 456–469. [CrossRef] [PubMed]
14. Bofill-Mas, S.; Albinana-Gimenez, N.; Clemente-Casares, P.; Hundesa, A.; Rodriguez-Manzano, J.; Allard, A.; Calvo, M.; Girones, R. Quantification and stability of human adenoviruses and polyomavirus JCPyV in wastewater matrices. *Appl. Environ. Microbiol.* **2006**, *72*, 7894–7896. [CrossRef] [PubMed]
15. Amarasiri, M.; Kawai, H.; Kitajima, M.; Okabe, S.; Sano, D. Specific interactions of rotavirus HAL1166 with Enterobacter cloacae SENG-6 and their contribution on rotavirus HAL1166 removal. *Water Sci. Technol.* **2019**, *79*, 342–348. [CrossRef] [PubMed]

16. Amarasiri, M.; Sano, D. Specific interactions between human norovirus and environmental matrices: Effects on the virus ecology. *Viruses* **2019**, *11*, 224. [CrossRef]
17. Miura, T.; Okabe, S.; Nakahara, Y.; Sano, D. Removal properties of human enteric viruses in a pilot-scale membrane bioreactor (MBR) process. *Water Res.* **2015**, *75*, 282–291. [CrossRef]
18. Amarasiri, M.; Kitajima, M.; Nguyen, T.H.; Okabe, S.; Sano, D. Bacteriophage removal efficiency as a validation and operational monitoring tool for virus reduction in wastewater reclamation: Review. *Water Res.* **2017**, *121*, 258–269. [CrossRef]
19. Ottoson, J.; Hansen, A.; Björlenius, B.; Norder, H.; Stenström, T.A. Removal of viruses, parasitic protozoa and microbial indicators in conventional and membrane processes in a wastewater pilot plant. *Water Res.* **2006**, *40*, 1449–1457. [CrossRef]
20. Marti, E.; Monclús, H.; Jofre, J.; Rodriguez-roda, I.; Comas, J.; Luis, J. Removal of microbial indicators from municipal wastewater by a membrane bioreactor (MBR). *Bioresour. Technol.* **2011**, *102*, 5004–5009. [CrossRef]
21. World Health Organization. *Guidelines for Drinking Water Quality: Fourth Edition Incorporating the First Addendum*, 4th ed.; World Health Organization: Geneva, Switzerland, 2017; ISBN 9789241549950.
22. Satter, S.M.; Aliabadi, N.; Gastañaduy, P.A.; Haque, W.; Mamun, A.; Flora, M.S.; Zaman, K.; Rahman, M.; Heffelfinger, J.D.; Luby, S.P.; et al. An update from hospital-based surveillance for rotavirus gastroenteritis among young children in Bangladesh, July 2012 to June 2017. *Vaccine* **2018**, *36*, 7811–7815. [CrossRef] [PubMed]
23. Tian, Y.; Chughtai, A.A.; Gao, Z.; Yan, H.; Chen, Y.; Liu, B.; Huo, D.; Jia, L.; Wang, Q.; MacIntyre, C.R. Prevalence and genotypes of group A rotavirus among outpatient children under five years old with diarrhea in Beijing, China, 2011–2016. *BMC Infect. Dis.* **2018**, *18*, 497. [CrossRef] [PubMed]
24. Kim, T.-D.; Unno, H. The roles of microbes in the removal and inactivation of viruses in a biological wastewater treatment system. *Water Sci. Technol.* **1996**, *33*, 243–250. [CrossRef]
25. Amarasiri, M.; Hashiba, S.; Miura, T.; Nakagomi, T.; Nakagomi, O.; Ishii, S.; Okabe, S.; Sano, D. Bacterial histo-blood group antigens contributing to genotype-dependent removal of human noroviruses with a microfiltration membrane. *Water Res.* **2016**, *95*, 383–391. [CrossRef] [PubMed]
26. Schmitz, B.W.; Kitajima, M.; Campillo, M.E.; Gerba, C.P.; Pepper, I.L. Virus reduction during advanced Bardenpho and conventional wastewater treatment processes. *Environ. Sci. Technol.* **2016**, *50*, 9524–9532. [CrossRef] [PubMed]
27. Arraj, A.; Bohatier, J.; Laveran, H.; Traore, O. Comparison of bacteriophage and enteric virus removal in pilot scale activated sludge plants. *J. Appl. Microbiol.* **2005**, *98*, 516–524. [CrossRef] [PubMed]
28. Zhang, C.-M.; Xu, L.-M.; Xu, P.-C.; Wang, X.C. Elimination of viruses from domestic wastewater: Requirements and technologies. *World J. Microbiol. Biotechnol.* **2016**, *32*, 69–77. [CrossRef] [PubMed]
29. Villabruna, N.; Koopmans, M.P.G.; De Graaf, M. Animals as reservoir for human norovirus. *Viruses* **2019**, *11*, 478. [CrossRef] [PubMed]
30. Summa, M.; Henttonen, H.; Maunula, L. Human noroviruses in the faeces of wild birds and rodents—New potential transmission routes. *Zoonoses Public Health* **2018**, *65*, 512–518. [CrossRef]
31. Uhrbrand, K.; Schultz, A.C.; Madsen, A.M. Exposure to airborne noroviruses and other bioaerosol components at a wastewater treatment plant in Denmark. *Food Environ. Virol.* **2011**, *3*, 130–137. [CrossRef]
32. Dueker, M.E.; O'Mullan, G.D.; Martinez, J.M.; Juhl, A.R.; Weathers, K.C. Onshore wind speed modulates Microbial aerosols along an urban waterfront. *Atmosphere* **2017**, *8*, 215. [CrossRef]
33. Ueki, Y.; Sano, D.; Watanabe, T.; Akiyama, K.; Omura, T. Norovirus pathway in water environment estimated by genetic analysis of strains from patients of gastroenteritis, sewage, treated wastewater, river water and oysters. *Water Res.* **2005**, *39*, 4271–4280. [CrossRef] [PubMed]
34. Kageyama, T.; Kojima, S.; Shinohara, M.; Uchida, K.; Fukushi, S.; Hoshino, F.B.; Takeda, N.; Katayama, K. Broadly reactive and highly sensitive assay for Norwalk-like viruses based on real-time quantitative reverse transcription-PCR. *J. Clin. Microbiol.* **2003**, *41*, 1548–1557. [CrossRef] [PubMed]
35. Amarasiri, M.; Kitajima, M.; Miyamura, A.; Santos, R.; Monteiro, S.; Miura, T.; Kazama, S.; Okabe, S.; Sano, D. Reverse transcription-quantitative PCR assays for genotype-specific detection of human noroviruses in clinical and environmental samples. *Int. J. Hyg. Environ. Health* **2018**, *221*, 578–585. [CrossRef] [PubMed]
36. Eftim, S.E.; Hong, T.; Soller, J.; Boehm, A.; Warren, I.; Ichida, A.; Nappier, S.P. Occurrence of norovirus in raw sewage—A systematic literature review and meta-analysis. *Water Res.* **2017**, *111*, 366–374. [CrossRef]

37. Wyn-Jones, A.P.; Carducci, A.; Cook, N.; D'Agostino, M.; Divizia, M.; Fleischer, J.; Gantzer, C.; Gawler, A.; Girones, R.; Höller, C.; et al. Surveillance of adenoviruses and noroviruses in European recreational waters. *Water Res.* **2011**, *45*, 1025–1038. [CrossRef] [PubMed]
38. Katayama, H.; Haramoto, E.; Oguma, K.; Yamashita, H.; Tajima, A.; Nakajima, H.; Ohgaki, S. One-year monthly quantitative survey of noroviruses, enteroviruses, and adenoviruses in wastewater collected from six plants in Japan. *Water Res.* **2008**, *42*, 1441–1448. [CrossRef]
39. Da Silva, A.K.; Le Saux, J.-C.; Parnaudeau, S.; Pommepuy, M.; Elimelech, M.; Le Guyader, F.S. Evaluation of removal of noroviruses during wastewater treatment, using real-time reverse transcription-PCR: Different behaviors of genogroups I and II. *Appl. Environ. Microbiol.* **2007**, *73*, 7891–7897. [CrossRef]
40. Da Silva, A.K.; Le Guyader, F.S.; Le Saux, J.C.; Pommepuy, M.; Montgomery, M.A.; Elimelech, M. Norovirus removal and particle association in a waste stabilization pond. *Environ. Sci. Technol.* **2008**, *42*, 9151–9157. [CrossRef]
41. Grassi, T.; Bagordo, F.; Idolo, A.; Lugoli, F.; Gabutti, G.; De Donno, A. Rotavirus detection in environmental water samples by tangential flow ultrafiltration and RT-nested PCR. *Environ. Monit. Assess.* **2010**, *164*, 199–205. [CrossRef]
42. Kiulia, N.; Hofstra, N.; Vermeulen, L.; Obara, M.; Medema, G.; Rose, J. Global occurrence and emission of rotaviruses to surface waters. *Pathogens* **2015**, *4*, 229–255. [CrossRef] [PubMed]
43. Baggi, F.; Peduzzi, R. Genotyping of rotaviruses in environmental water and stool samples in Southern Switzerland by nucleotide sequence analysis of 189 base pairs at the 5′ end of the VP7 gene. *J. Clin. Microbiol.* **2000**, *38*, 3681–3685. [PubMed]
44. Tree, J.A.; Adams, M.R.; Lees, D.N. Chlorination of indicator bacteria and viruses in primary sewage effluent. *Appl. Environ. Microbiol.* **2003**, *69*, 2038–2043. [CrossRef] [PubMed]
45. Albinana-Gimenez, N.; Miagostovich, M.P.; Calgua, B.; Huguet, J.M.; Matia, L.; Girones, R. Analysis of adenoviruses and polyomaviruses quantified by qPCR as indicators of water quality in source and drinking-water treatment plants. *Water Res.* **2009**, *43*, 2011–2019. [CrossRef]
46. Haramoto, E.; Fujino, S.; Otagiri, M. Distinct behaviors of infectious F-specific RNA coliphage genogroups at a wastewater treatment plant. *Sci. Total Environ.* **2015**, *520*, 32–38. [CrossRef] [PubMed]
47. Kazama, S.; Masago, Y.; Tohma, K.; Souma, N.; Imagawa, T.; Suzuki, A.; Liu, X.; Saito, M.; Oshitani, H.; Omura, T. Temporal dynamics of norovirus determined through monitoring of municipal wastewater by pyrosequencing and virological surveillance of gastroenteritis cases. *Water Res.* **2016**, *92*, 244–253. [CrossRef]
48. Kazama, S.; Miura, T.; Masago, Y.; Konta, Y.; Tohma, K.; Manaka, T.; Liu, X.; Nakayama, D.; Tanno, T.; Saito, M.; et al. Environmental Surveillance of norovirus genogroups I and II for sensitive detection of epidemic variants. *Appl. Environ. Microbiol.* **2017**, *83*, e03406-16. [CrossRef] [PubMed]
49. Haramoto, E.; Kitajima, M.; Hata, A.; Torrey, J.R.; Masago, Y.; Sano, D.; Katayama, H. A review on recent progress in the detection methods and prevalence of human enteric viruses in water. *Water Res.* **2018**, *135*, 168–186. [CrossRef]
50. Mijatovic-rustempasic, S.; Esona, M.D.; Williams, A.L.; Bowen, M.D. Sensitive and specific nested PCR assay for detection of rotavirus A in samples with a low viral load. *J. Virol. Methods* **2016**, *236*, 41–46. [CrossRef]
51. Lew, A.M.; Marshall, V.M.; Kemp, D.J. Affinity selection of polymerase chain reaction products by DNA-binding proteins. In *Methods in Enzymology*; Academic Press: Cambridge, MA, USA, 1993; Volume 218, pp. 526–534.
52. Ma, X.Y.; Wang, X.C.; Wang, D.; Ngo, H.H.; Zhang, Q.; Wang, Y.; Dai, D. Function of a landscape lake in the reduction of biotoxicity related to trace organic chemicals from reclaimed water. *J. Hazard. Mater.* **2016**, *318*, 663–670. [CrossRef]
53. Gao, T.; Chen, R.; Wang, X.; Hao, H.; Li, Y.; Zhou, J.; Zhang, L. Application of disease burden to quantitative assessment of health hazards for a decentralized water reuse system. *Sci. Total Environ.* **2016**, *551–552*, 83–91. [CrossRef] [PubMed]
54. Lewis, G.D.; Metcalf, T.G. Polyethylene glycol precipitation for recovery of pathogenic viruses, including hepatitis A virus and human rotavirus, from oyster, water, and sediment samples. *Appl. Environ. Microbiol.* **1988**, *54*, 1983–1988. [PubMed]
55. Ji, Z.; Wang, X.; Zhang, C.; Miura, T.; Sano, D.; Funamizu, N.; Okabe, S. Occurrence of hand-foot-and-mouth disease pathogens in domestic sewage and secondary effluent in Xi'an, China. *Microbes Environ.* **2012**, *27*, 288–292. [CrossRef] [PubMed]

56. Katayama, H.; Shimasaki, A.; Ohgaki, S. Development of a virus concentration method and its application to detection of enterovirus and Norwalk virus from coastal seawater. *Appl. Environ. Microbiol.* **2002**, *68*, 1033–1039. [CrossRef]
57. O'Neill, H.J.; McCaughey, C.; Coyle, P.V.; Wyatt, D.E.; Mitchell, F. Clinical utility of nested multiplex RT-PCR for group F adenovirus, rotavirus and norwalk-like viruses in acute viral gastroenteritis in children and adults. *J. Clin. Virol.* **2002**, *25*, 335–343. [CrossRef]
58. Gilgen, M.; Germann, D.; Lüthy, J.; Hübner, P. Three-step isolation method for sensitive detection of enterovirus, rotavirus, hepatitis A virus, and small round structured viruses in water samples. *Int. J. Food Microbiol.* **1997**, *37*, 189–199. [CrossRef]
59. Kojima, S.; Kageyama, T.; Fukushi, S.; Hoshino, F.B.; Shinohara, M.; Uchida, K.; Natori, K.; Takeda, N.; Katayama, K. Genogroup-specific PCR primers for detection of Norwalk-like viruses. *J. Virol. Methods* **2002**, *100*, 107–114. [CrossRef]
60. Kumar, S.; Stecher, G.; Tamura, K. MEGA7: Molecular Evolutionary Genetics Analysis version 7.0 for bigger datasets. *Mol. Biol. Evol.* **2016**, *33*, 1870–1874. [CrossRef]
61. Kimura, M. A simple method for estimating evolutionary rates of base substitutions through comparative studies of nucleotide sequences. *J. Mol. Evol.* **1980**, *16*, 111–120. [CrossRef]
62. Felsenstein, J. Confidence limits on phylogenies: An approach using the bootstrap. *Evolution (N. Y.)* **1985**, *39*, 783–791.
63. Letunic, I.; Bork, P. Interactive Tree Of Life (iTOL) v4: Recent updates and new developments. *Nucleic Acids Res.* **2019**, *47*, W256–W259. [CrossRef] [PubMed]

pathogens

MDPI

Article

Metagenomic Analysis of Infectious F-Specific RNA Bacteriophage Strains in Wastewater Treatment and Disinfection Processes

Suntae Lee *, Mamoru Suwa and Hiroyuki Shigemura

Innovative Materials and Resources Research Center, Public Works Research Institute, Ibaraki 305-8516, Japan; suwa@pwri.go.jp (M.S.); h-shigemura@pwri.go.jp (H.S.)
* Correspondence: seontae@pwri.go.jp

Received: 24 September 2019; Accepted: 2 November 2019; Published: 3 November 2019

Abstract: F-specific RNA bacteriophages (FRNAPHs) can be used to indicate water contamination and the fate of viruses in wastewater treatment plants (WWTPs). However, the occurrence of FRNAPH strains in WWTPs is relatively unknown, whereas FRNAPH genotypes (GI–GIV) are well documented. This study investigated the diversity of infectious FRNAPH strains in wastewater treatment and disinfection processes using cell culture combined with next-generation sequencing (integrated culture–NGS (IC–NGS)). A total of 32 infectious strains belonging to FRNAPH GI (nine strains), GI-JS (two strains), GII (nine strains), GIII (seven strains), and GIV (five strains) were detected in wastewater samples. The strains of FRNAPH GI and GII exhibited greater resistance to wastewater treatment than those of GIII. The IC–NGS results in the disinfected samples successfully reflected the infectivity of FRNAPHs by evaluating the relationship between IC–NGS results and the integrated culture–reverse-transcription polymerase chain reaction combined with the most probable number assay, which can detect infectious FRNAPH genotypes. The diversity of infectious FRNAPH strains in the disinfected samples indicates that certain strains are more resistant to chlorine (DL52, GI-JS; T72, GII) and ultraviolet (T72, GII) disinfection. It is possible that investigating these disinfectant-resistant strains could reveal effective mechanisms of viral disinfection.

Keywords: F-specific RNA bacteriophage strain; viral indicator; next-generation sequencing; infectivity; wastewater treatment; chlorination; ultraviolet disinfection

---

1. Introduction

F-specific RNA bacteriophages (FRNAPHs), which are known to infect *Escherichia coli* that express F pili, have a single-stranded RNA genome enclosed in an icosahedral capsid measuring 20–30 nm in diameter. The sizes, shape structures, and genomes of FRNAPHs are similar to those of noroviruses [1,2], which have caused numerous outbreaks of gastroenteritis in multiple countries [3]. Furthermore, FRNAPH behavior, abundance, and survival in the environment including during water treatment are also similar to those of human enteric viruses [1,2,4–6]. Thus, they serve as potential indicators of water contamination and the fates of viruses in aquatic environments and wastewater treatment plants (WWTPs) [4–6].

FRNAPHs belong to the family *Leviviridae* and are classified into the genera *Levivirus* and *Allolevivirus*, which are subdivided into genotypes I and II (GI and GII) and genotypes III and IV (GIII and GIV), respectively. Each FRNAPH genotype has a different fate in WWTPs [6–10] and a different resistance to disinfection [11–13]. For example, genotypes GII and GIII are more prevalent than GI and GIV in municipal raw wastewater samples [6,8,9]. However, GI is the dominant genotype in the secondary effluent of WWTPs because of its higher resistance to wastewater treatment relative to other FRNAPH genotypes [6,8,9]. GI also showed the highest chlorine and ultraviolet resistance among the

FRNAPH genotypes [11–13]. Particularly, MS2, belonging to GI, showed higher ultraviolet resistance than human pathogenic viruses (poliovirus, rotavirus, hepatitis A virus, and coxsackievirus) [13].

The presence and removal of FRNAPH genotypes in WWTPs have been the subject of numerous studies [6–10]. Moreover, several FRNAPH strains are included in each FRNAPH genotype and have been reported worldwide in bacterial isolates associated with sewage and mammal feces [14–18]. The sources from where different FRNAPH strains were first isolated/detected are shown in Table 1. In the last decade, novel GI-JS strains DL52 and DL54 were isolated, which are recombinant strains of environmental isolates of *Leviviridae* ssRNA bacteriophages [16]. Unfortunately, information regarding the occurrence of FRNAPH strains in WWTPs is relatively limited [19].

Table 1. Sources of F-specific RNA bacteriophage (FRNAPH) strains.

FRNAPH Genotype	FRNAPH Strain	Source	Reference
GI	MS2	Sewage	[14–16]
	M12	Sewage	[14–16]
	DL1	River water	[14–16]
	DL2	Bay water	[14–16]
	DL13	Oyster	[14–16]
	DL16	Bay water	[14–16]
	J20	Chicken litter	[14–16]
	ST4	Unknown	[14–16]
	R17	Sewage	[14–16]
	Fr	Dung hill	[14,16]
	JP501	Sewage	[17]
GI-JS	DL52	Bay water	[16]
	DL54	Bay water	[16]
GII	GA	Sewage	[14–17]
	KU1	Sewage	[14–17]
	DL10	Mussel	[14–16]
	DL20	Clam	[14–16]
	T72	Bird	[14–16]
	BZ13	Sewage	[17]
	TL2	Sewage	[17]
	JP34	Sewage	[17]
	TH1	Sewage	[17]
GIII	Qβ	Human feces	[14,15,17]
	BR12	Creek water	[14,15]
	BZ1	Sewage	[14,15]
	VK	Sewage	[14,15,17]
	TW18	Sewage	[14,15,17]
	HL4-9	Hog lagoon	[14,15]
	M11	Unknown	[14,15]
	MX1	Sewage	[14,15,17]
GIV	SP	Siamang gibbon	[14,15,17,18]
	FI	Infant	[14,15,17,18]
	BR1	Creek water	[14,15]
	BR8	Creek water	[14,15]
	HB-P22	Bird	[14,15]
	HB-P24	Bird	[14,15]
	NL95	Calf	[14,15]

Numerous studies have employed MS2, GA, Qβ, and SP as representative FRNAPH strains of genotypes GI–GIV in spiking experiments to determine their surface properties, including electrostatic surface charge, hydrophobicity, and removal during water treatment processes such as coagulation and membrane filtration [20–22]. However, the dominance of these strains among the strains of each FRNAPH genotype is debated. Thus, it is particularly important to identify the dominant strains affecting the concentrations of FRNAPH genotypes.

FRNAPH GI and GIV predominantly occur in the feces and waste generated by animal farms, whereas FRNAPH GII and GIII are dominant in human feces and the raw sewage of WWTPs [23,24].

Thus, the distribution of FRNAPH genotypes has been widely studied in order to determine the source of fecal contamination in river water [7,25–27], shellfish [26,28,29], and sediments [27]. However, a previous study [19] suggested that this distribution is not sufficient for tracking the source of fecal pollution. The large diversity of FRNAPH strains in each genotype may be the reason for this limitation because they are found in a diverse range of water bodies (e.g., sewage, river water, and seawater), shellfish (oysters, mussels, and clams), and the feces of birds and mammals (including humans, chicken, swine, calves, and apes) [14–18]. For example, FRNAPH GI strains MS2, DL1, and J20 have been isolated from wastewater, river water, and chicken litter, respectively (Table 1). Therefore, it is important to investigate the diversity of FRNAPH strains.

Next-generation sequencing (NGS) is used to study viral metagenomes in different stages of wastewater treatment [30–33]. This method provides more conservative estimates of viral occurrence compared with the rates detected using quantitative polymerase chain reaction (qPCR) [32]. The advantage of metagenomics is that it allows a comprehensive characterization of FRNAPH strain diversity. However, like qPCR assays, metagenomic methods do not assess infectivity. Therefore, when samples acquired after disinfection using chlorine or ultraviolet light are subjected to NGS, the viral sequences do not reflect infectivity. Conversely, culture combined with PCR (integrated culture–PCR (IC–PCR)) can detect infectious viruses. For example, IC–RT-PCR combined with a most probable number (MPN) assay (IC–RT-PCR–MPN) has been used to quantitatively detect infectious FRNAPH genotypes [6,34,35]. Thus, we hypothesized that the application of NGS for detecting of FRNAPH strains propagated in a liquid medium may be effective for detecting infectious FRNAPH strains. NGS analyses of wastewater samples often show that the majority of genes are from eukaryotes and bacteria, which are more abundant than viruses and bacteriophages. However, propagating infectious FRNAPH strains in samples can result in large yields of FRNAPH sequences; it also differentiates between infective and inactive FRNAPH strains. Recently, known and novel plant viruses, which infect plants such as yams, were detected by NGS combined with robust yam propagation by tissue-culture [36]. NGS combined with cell culture was also used to characterize enteric viruses isolated from wastewater [33]. Thus, integrated culture–NGS (IC–NGS) can be used to detect infectious FRNAPH strains and high fractions of FRNAPH genes in wastewater samples.

To the best of our knowledge, this study is the first to use IC–NGS to investigate the diversity of infectious FRNAPH strains in wastewater treatment and disinfection processes. We prepared the influent and secondary effluent of a WWTP as well as disinfected secondary effluent (raw water) treated using chlorine or ultraviolet light. IC–NGS and IC–RT-PCR–MPN were performed to determine the diversity of infectious FRNAPH strains and the concentrations of infectious FRNAPH genotypes, respectively. The relationship between the results of the two assays was investigated to evaluate whether IC–NGS data can effectively reflect the infectivity of FRNAPHs.

2. Results

2.1. Metagenomic and Taxonomic Analyses

A summary of the metagenomic (BLASTn) and taxonomic (MEGAN) analyses is shown in Table 2. The numbers of reads of the 12 samples analyzed using IC–NGS ranged from 887,593 to 5,035,503, and the trimmed sequences were assembled into 611–18,941 contigs. The FRNAPH strains were represented in the contigs of all samples using IC–NGS, and 66–551 sequences represented the reference genomes of FRNAPH strains determined using BLASTn. The percentages of hits for FRNAPH strains relative to the number of contigs in the samples ranged from 3% to 36%. The vast majority of the hit sequences assigned using MEGAN represented bacterial sequences and ranged from 44% to 83%. Specifically, *Salmonella enterica* sequences dominated in the bacterial sequences (65–92% without 1127 influent sample). The range of contigs that did not correspond to a reference genome was 4–36%.

Table 2. Characteristics of influent and secondary effluent samples [1].

Date (Month/Day)	Sample [2]	No. of Total Reads	No. of Contigs	No. of Hits for FRNAPHs (Ratio)	No. of Hits for Bacteria (Ratio), [No. of Hits for *Salmonella enterica* (ratio)] [3]	No. of not Hit Contigs (Ratio) [4]
11/13	IN	1,135,519	1218	317 (26%)	584 (48%), [380 (65%)]	200 (16%)
	SE	1,080,326	537	87 (16%)	343 (64%), [261 (76%)]	66 (12%)
	Cl	887,593	611	73 (12%)	476 (78%), [414 (87%)]	30 (5%)
	UV	1,278,120	732	66 (9%)	608 (83%), [548 (90%)]	41 (6%)
11/20	IN	1,070,341	1299	468 (36%)	570 (44%), [459 (81%)]	196 (15%)
	SE	1,019,493	614	160 (26%)	310 (50%), [220 (71%)]	95 (15%)
	Cl	1,033,979	776	91 (12%)	591 (76%), [532 (90%)]	52 (7%)
	UV	1,025,377	821	162 (20%)	577 (70%), [505 (88%)]	36 (4%)
11/27	IN	4,092,357	18,941	551 (3%)	10,471 (55%), [2521 (24%)]	6859 (36%)
	SE	4,900,897	4344	247 (6%)	2537 (58%), [1825 (72%)]	1151 (26%)
	Cl	5,035,503	4370	161 (4%)	3484 (80%), [3217 (92%)]	497 (11%)
	UV	4,102,143	2319	106 (5%)	1655 (71%), [1416 (86%)]	497 (21%)

[1] The number of hits for each FRNAPH or bacterial genome refers to the number of sequences registering hits for FRNAPH genomes or bacterial reference genomes. The ratio is the percentage of the number of hits relative to the number of total contigs in the sample. [2] IN: Influent; SE: Secondary effluent; Cl: Chlorine-treated secondary effluent samples; UV: Ultraviolet-treated secondary effluent samples. [3] The ratio shown for *Salmonella enterica* is the number of hits for *Salmonella enterica* relative to the number for all bacteria. [4] Not hit contigs refers to the absence of hits for any reference genome.

2.2. Detection of Infectious FRNAPH Strains in Wastewater Treatment and Disinfection Processes

IC–NGS detected 31 stains representing all FRNAPH genotypes in influent, secondary effluent, chlorine-treated, and ultraviolet-treated samples on 11/13, 11/20, and 11/27 (Figure 1). The GI strains MS2, DL1, J20, fr, DL16, JP501, R17, ST4, and M12 were detected in all 12 samples (Figure 1). Specifically, MS2, DL1, and J20 were the most frequently detected GI strains (12/12, 100%). The proportions of GI strains in the secondary effluent samples were higher than those in the influent samples. The proportions of abundant GI strains (MS2, DL1, and J20) decreased from secondary effluent samples to chlorine- and ultraviolet-treated samples.

DL52 and DL54 (FRNAPH GI-JS) were detected in all samples (Figure 1). DL52 was the predominant strain of FRNAPHs in the influent samples together with HL4-9 (FRNAPH GIII). The proportions of DL52 decreased to a greater extent from influent to secondary effluent samples than those of DL54. In contrast, the proportions of DL54 decreased compared to those of DL52 from secondary effluent samples to chlorine-treated and ultraviolet-treated samples. The proportions of DL52 in chlorine-treated and ultraviolet-treated samples were similar or higher than those in the secondary effluent samples; specifically, the proportion of DL52 in the chlorine-treated sample from 11/20 (26.4%) was the highest among all FRNAPH strains.

The FRNAPH GII strains DL20, T72, GA, DL10, JP34, KU1, BZ13, TL2, and TH1 were detected in all 12 samples (Figure 1). Moreover, DL20 was the most predominant strain of FRNAPH GII in influent and secondary effluent samples (34.2–48.5% and 30.0–57.1% of the proportions in GII genotypes, respectively, Supplementary Figure S2). The proportions of GII strains in secondary effluent samples were higher than those in influent samples. DL20 had the highest proportion of all strains in the secondary effluent sample from 11/13 (18.4%). Furthermore, the proportions of FRNAPH GII strains in the chlorine-treated and ultraviolet-treated samples were similar or higher than those in the secondary effluent samples. Specifically, DL20 and T72 had the highest proportion in chlorine-treated and ultraviolet-treated samples from 11/13 (24.7% and 25.8%, respectively) and in chlorine-treated samples from 11/27 (23.6%) and ultraviolet-treated samples from 11/20 (27.8%), respectively.

The FRNAPH GIII strains HL4-9, Qβ, TW18, VK, BR12, BZ1, and M11 were detected in all 12 samples (Figure 1). HL4-9, which was detected in all samples (12/12, 100%), was the most abundant strain of FRNAPH GIII in all samples except chlorine-treated samples from 11/13 (28.6–83.3% of GIII genotypes, Supplementary Figure S2). Moreover, all FRNAPH strains in the influent samples together with DL52 represent FRNAPH GI-JS. The proportions of all strains of FRNAPH GIII in the influent samples was reduced by wastewater treatment (secondary effluent samples) and by chlorine (chlorine-treated samples) and ultraviolet disinfection (ultraviolet-treated samples), with the exception of ultraviolet-treated samples collected on 11/27 for HL4-9.

The FRNAPH GIV strains FI, BR1, BR8, HB-P22, and SP were detected in all 12 samples (Figure 1). FI and BR1 were the predominant FRNAPH GIV strains in all samples (Supplementary Figure S2). SP was detected only once in the ultraviolet-treated samples from 11/13. The proportion of FI increased to a greater extent from influent to secondary effluent samples on 11/13 compared to those on other dates, which were either similar or smaller. There were fewer hits for GIV strains in chlorine-treated and ultraviolet-treated samples (<9, Supplementary Figure S1).

FRNAPH genotype	FRNAPH strain	No. positive/ no. tested (%)	Proportions of FRNAPH strains (%)											
			IN			SE			Cl			UV		
			11/13	11/20	11/27	11/13	11/20	11/27	11/13	11/20	11/27	11/13	11/20	11/27
GI	MS2	12/12 (100)	2.8	3.0	2.9	2.3	7.5	5.3	2.7	1.1	4.3	1.5	2.5	5.7
	DL1	12/12 (100)	3.8	1.1	3.6	2.3	5.6	8.9	2.7	1.1	2.5	3.0	1.9	7.5
	J20	12/12 (100)	1.3	2.4	3.1	5.7	4.4	8.9	2.7	2.2	3.1	1.5	1.2	5.7
	fr	10/12 (83)	0.6	0.9	2.7		1.9	0.8	2.7	6.6	0.6	3.0	3.1	
	DL16	9/12 (75)	1.3	1.3	0.7	2.3	3.8	4.5	1.4		3.1			6.6
	JP501	5/12 (42)		0.6	0.9		0.6	0.4				4.5		
	R17	5/12 (42)		0.2	0.4		5.0	1.2						0.9
	ST4	4/12 (33)		0.4	0.2		1.3					1.5		
	M12	3/12 (25)		0.4			0.6			1.1				
GI-JS	DL52	12/12 (100)	14.5	16.2	20.0	6.9	9.4	9.7	6.8	26.4	10.6	6.1	11.7	10.4
	DL54	12/12 (100)	7.3	5.3	5.6	4.6	5.6	12.6	4.1	5.5	3.1	6.1	2.5	9.4
GII	DL20	12/12 (100)	4.1	5.3	5.8	18.4	10.0	4.9	24.7	15.4	17.4	25.8	21.0	8.5
	T72	12/12 (100)	0.9	1.9	1.5	5.7	2.5	3.2	13.7	3.3	23.6	1.5	27.8	7.5
	GA	12/12 (100)	3.2	1.7	2.2	8.0	1.3	2.8	12.3	2.2	3.1	6.1	4.9	2.8
	DL10	12/12 (100)	0.3	1.5	0.5	2.3	1.9	0.8	4.1	4.4	6.2	9.1	8.6	6.6
	JP34	9/12 (75)	1.9	0.6	0.9		0.6	2.4	5.5		3.1	4.5	1.2	
	KU1	8/12 (67)	0.3	0.6	0.4	1.1		0.8		2.2	1.9		4.3	
	BZ13	6/12 (50)	1.3				1.3	0.8		2.2		1.5	1.2	
	TL2	5/12 (42)			0.5	1.1			1.4		0.6		1.2	
	TH1	3/12 (25)			0.2			0.4					0.6	
GIII	HL4-9	12/12 (100)	18.0	17.3	17.2	14.9	13.8	13.4	1.4	2.2	8.7	7.6	3.1	17.9
	Qβ	11/12 (92)	12.0	9.8	9.6	3.4	6.9	4.5	1.4	2.2	0.6		0.6	2.8
	TW18	11/12 (92)	7.6	6.4	5.1	1.1	8.1	3.6	4.1	2.2	0.6	1.5		2.8
	VK	8/12 (67)	7.3	8.5	4.9		1.3	0.8			1.2	3.0		1.9
	BR12	7/12 (58)	5.4	7.3	5.6		0.6	2.8		1.1		1.5		
	BZ1	5/12 (42)	0.9	2.6	1.5		0.6	1.6						
	M11	4/12 (33)	0.3	0.9	0.2		1.9							
GIV	FI	12/12 (100)	3.8	3.0	1.6	17.2	3.1	0.8	5.5	8.8	1.9	3.0	1.9	0.9
	BR1	12/12 (100)	0.9	0.2	1.5	1.1	0.6	4.0	2.7	8.8	3.7	6.1	0.6	1.9
	BR8	5/12 (42)	0.3	0.2	0.4	1.1				1.1				
	HB-P22	2/12 (17)		0.2	0.4									
	SP	1/12 (8)										1.5		

Figure 1. Proportions of FRNAPH strains representing each genotype in 12 samples combined with a heat map showing the relative abundance of all FRNAPH strains according to the number of hits (Supplementary Figure S1) in the BLASTn analyses of influent (IN), secondary effluent (SE), chlorine-treated (Cl), and ultraviolet-treated (UV) samples. Proportions (%) for FRNAPH strains were calculated as the number of hits for a specific FRNAPH strain relative to the total hits for all FRNAPH strains in each sample. Blank cells indicate an absence of hits. Green and red cells indicate the lowest and highest values, respectively. Numbers in the heat-map cells indicate the proportions for samples collected on 11/13, 11/20, and 11/27.

2.3. Comparison of IC–RT-PCR–MPN and IC–NGS Data

The relationship between the results for infectious FRNAPH genotypes detected using IC–RT-PCR–MPN and IC–NGS was investigated to determine whether IC–NGS effectively reflects the infectivity of FRNAPHs. In the IC–RT-PCR–MPN results (Figure 2A), infectious FRNAPH GII was detected in all chlorine-treated samples, whereas GI was not detected. GIII and GIV were detected in chlorine-treated samples collected on 11/20 and 11/27 and 11/13 and 11/20, respectively. GI and GIII were inactivated more effectively by chlorine disinfection (GI, >1.6 to >3.7 $\log_{10}$; GIII, 1.4 to >3.2 $\log_{10}$) than GII and GIV. After ultraviolet disinfection, infectious FRNAPH GII was detected in all ultraviolet-treated samples, whereas GIII was not detected. GI and GIV were detected in ultraviolet-treated samples collected on 11/27 and 11/13, respectively. The highest inactivation among all infectious FRNAPH genotypes was observed for GIII (>2.4–>3.2 $\log_{10}$).

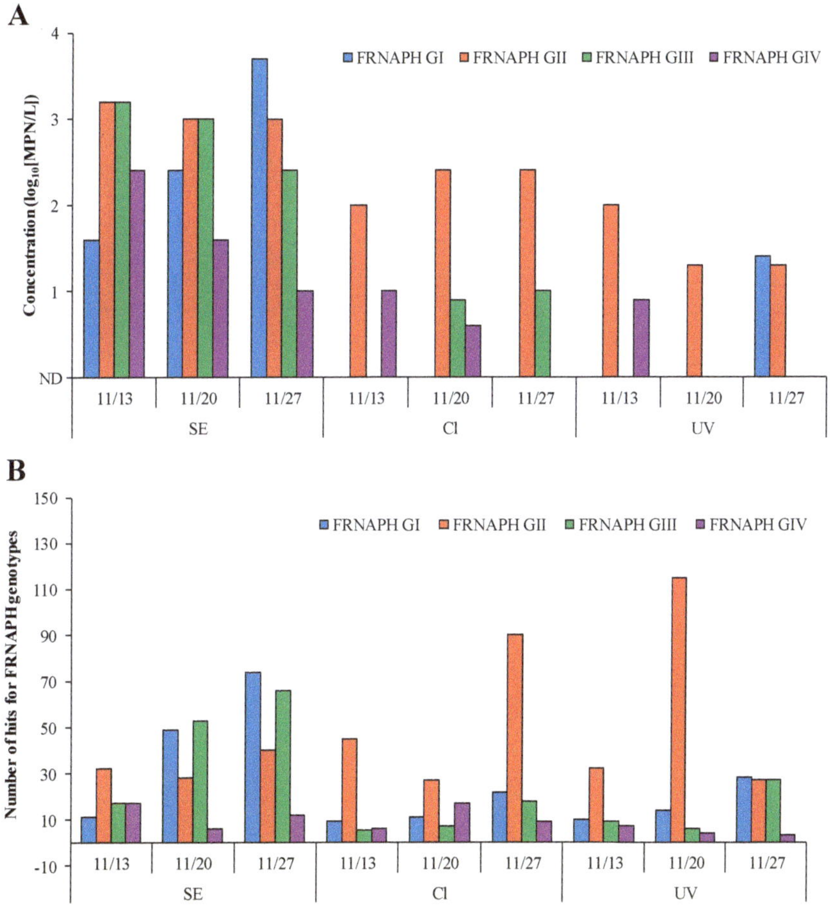

Figure 2. Concentrations of infectious FRNAPH genotypes determined using IC–RT-PCR–MPN (**A**) and number of hits for each FRNAPH genotype determined using IC–NGS (**B**) in the secondary effluent (SE), chlorine-treated (Cl), and ultraviolet-treated (UV) samples collected on 11/13, 11/20, and 11/27. Numbers of hits for each FRNAPH genotype represent the sum of the number of hits for FRNAPH strains of each genotype except GI-JS.

Figure 2B shows the number of hits for each FRNAPH genotype except GI-JS from the sum of the number of hits for each genotype (Supplementary Figure S1) in the secondary effluent, chlorine-treated, and ultraviolet-treated samples collected on 11/13, 11/20, and 11/27. We observed the highest ratio of hits for GII (27–115) among the FRNAPH genotypes from chlorine-treated and ultraviolet-treated samples. In particular, 90 and 115 hits were observed in the 11/27 chlorine-treated sample and the 11/20 ultraviolet-treated sample, respectively. Notably, only GII was detected using IC–RT-PCR–MPN. In contrast, the largest decreases among the infectious FRNAPH genotypes were observed among GI and GIII strains in the chlorine-treated samples (18–87% decrease) and GIII strains in the ultraviolet-treated samples (18–87% decrease). These trends were equivalent to those observed using IC–RT-PCR–MPN.

3. Discussion

The aim of this study was to investigate the diversity of infectious FRNAPH strains in wastewater treatment and disinfection processes using IC–NGS. A total of 32 FRNAPH strains were successfully detected in wastewater samples by IC–NGS (Figure 1). These strains have been first isolated from various sources that include not only sewage and environmental waters but also shellfish and human and animal feces (Table 1). This indicates that multiple FRNAPH strains from various sources accumulate in the influent of WWTPs. DL52 (GI-JS) and HL4-9 (GIII) were predominant, representing more than 30% of FRNAPH strains identified in influent samples from the target WWTP. DL52 and HL4-9 were first isolated from bay water and hog lagoons, respectively (Table 1). HL4-9 has been associated with pig waste. The results of our previous study [6], which investigated the occurrence of FRNAPH genotypes in the same WWTP, suggested that livestock waste was present in the influent. Thus, wastewater related to pig farming may be incorporated into the influent of the target WWTP in this study as well. These results suggest that the identification of FRNAPH strains by IC–NGS could be useful for microbial source tracking; however, further investigation is required to identify infectious FRNAPH strains from more specific sources such as animal feces and abattoir wastewater.

A comparison of the proportions of dominant DL52 and HL4-9 in influent and secondary effluent samples revealed that DL52 decreased to a greater extent than HL4-9 (Figure 1). This indicates that DL52 was more efficiently removed by wastewater treatment than HL4-9. Furthermore, the proportions of most strains of FRNAPH GI and GII were similar or higher in the secondary effluent relative to those in the influent, whereas those of GIII, including HL4-9, were decreased. This suggests that strains of FRNAPH GI and GII are more resistant to wastewater treatment than those of GIII. Previous studies determined by IC–RT-PCR–MPN and RT-qPCR have also shown smaller reductions of GI and GII by wastewater treatment when compared to GIII [6,9]. Thus, the results of this study determined by IC–NGS agree with those of previous research. Conversely, DL52 and DL54, which belong to the same genotype (GI-JS), showed different proportions in influent and secondary effluent samples. The proportions of DL54 in secondary effluent samples collected on 11/20 and 11/27 were similar or higher than those in influent samples, while those of DL52 were significantly lower. This result suggests differences in wastewater treatment efficacy for different strains of the same genotype. However, further investigation is required using RT-qPCR in order to evaluate the removal quantities for each strain.

After chlorine disinfection (Figure 1, Cl), DL20 (GII), DL52 (GI-JS), and T72 (GII) were predominant with >20% for 11/13, 11/20, and 11/27 samples, respectively. In particular, DL52 and T72 were not predominant before chlorination in 11/20 and 11/27 samples (Figure 1, SE), whereas DL20 was predominant in the secondary effluent sample collected on 11/13. This indicates that DL52 and T72 is more resistant to chlorination than other FRNAPH strains. Similarly, whereas DL20 (GII) and HL4-9 (GIII) were predominant before and after ultraviolet disinfection in 11/13 and 11/27 samples (Figure 1, SE and UV), respectively, T72 (GII) was only predominant in the ultraviolet-treated sample collected on 11/20. This also indicates that T72 may be more resistant to ultraviolet disinfection than other FRNAPH strains. Future research should confirm the disinfectant resistance of these strains (DL52 and T72) through experiments using isolates of these strains.

Previous studies of the surface properties and removal of FRNAPH genotypes during water treatment used MS2, GA, Qβ, and SP as representative FRNAPH strains of genotypes GI–GIV [20–22]. However, DL20, HL4-9, and FI were more predominant strains of FRNAPH GII, GIII, and GIV in our wastewater samples than GA, Qβ, and SP, respectively. Specifically, SP, which was detected only once (ultraviolet-treated sample collected on 11/13), was rarely found in the wastewater samples. Thus, our results suggest that DL20, HL4-9, and FI are more representative FRNAPH strains of genotypes GII, GIII, and GIV, respectively.

One of the objectives of this study was to evaluate whether IC–NGS data can effectively reflect the infectivity of FRNAPHs by comparing the detection of infectious FRNAPH genotypes using IC–RT-PCR–MPN and IC–NGS. It should be noted that FRNAPH GII showed a higher concentration and number of hits than FRNAPH genotypes GI, GIII, and GIV when IC–RT-PCR–MPN and IC–NGS were used to analyze chlorine-treated and ultraviolet-treated samples (Figure 2A,B). Further, the largest decreases in the number of hits among all infectious FRNAPH genotypes were observed for GI and GIII strains from secondary effluent to chlorine-treated samples as well as GIII strains from secondary effluent to ultraviolet-treated samples (Figure 2B). These data are consistent with those acquired using IC–RT-PCR–MPN (Figure 2A). These results indicate that the infectivity of FRNAPHs is reflected by the IC–NGS data when infectious FRNAPHs are propagated before performing NGS.

Viral diversity measured by NGS varies among studies because of pre-treatment processes such as nucleic-acid extraction and inherent amplification biases during PCR [37,38]. In the IC–NGS results, specific strains that easily propagated during the pre-propagating procedure prior to NGS were more frequently detected by IC–NGS. If specific strains are easily propagated, the distributions of FRNAPH strains would be similar in all samples. However, the distributions of FRNAPH strains differed between influent, secondary effluent, chlorine-treated, and ultraviolet-treated samples, and between those collected on 11/13, 11/20, and 11/27, except for the influent sample (Figure 1). Thus, the propagating bias may not have affected the results of this study. On the other hand, the distribution of FRNAPH strains may have been affected by the culture conditions (temperature, culture time, using the host strain, etc.) in the pre-propagating procedure of IC–NGS. Thus, further studies are needed to investigate the effect of the culture conditions used for IC–NGS on the distribution of FRNAPH strains.

In conclusion, this study revealed that diverse infectious FRNAPH strains are present in wastewater treatment and disinfection processes by IC–NGS. A total of 32 infectious strains belonging to FRNAPH GI (nine strains), GI-JS (two strains), GII (nine strains), GIII (seven strains), and GIV (five strains) were detected in the wastewater samples from a pilot-scale WWTP. The GI and GII strains were more resistant to wastewater treatment than GIII strains. The IC–NGS results from disinfected samples reflected the infectivity of FRNAPHs. Our results suggest that certain strains exhibit greater resistance to chlorine (DL52, GI-JS; T72, GII) and ultraviolet (T72, GII) disinfection than others from the results of laboratory-scale batch disinfection experiments, using secondary effluent samples. The results of this study will be confirmed by investigating full-scale WWTPs. By identifying disinfectant-resistant strains, it is likely that further research will reveal more effective mechanisms for viral disinfection, thereby reducing viruses at WWTPs for ensuring the hygiene and safety of recreational waters.

4. Materials and Methods

4.1. Wastewater Samples

Influent and secondary effluent samples were collected from a pilot-scale WWTP (capacity of 10 m^3/d), which uses conventional activated sludge treatment with 1700–2100 mg/L of mixed-liquor suspended solids. This WWTP is fed by water from the influent of a full-scale WWTP located in Ibaraki Prefecture, Japan. The influent and secondary effluent samples were collected on November 13, 20, and 27, 2017 (designated 11/13, 11/20, and 11/27, respectively). The characteristics of the influent and secondary effluent samples are summarized in Table 3.

Table 3. Characteristics of influent and secondary effluent samples.

Parameter [1]	Units	Range	
		IN [2]	SE [2]
pH	-	7.1–7.3	6.8–6.9
CODcr	mg/L	120–140	11–14
SS	mg/L	47–78	4.7–6.7
Turbidity	NTU	37–44	1.2–2.8
T-N	mg/L	31–34	15–17
T-P	mg/L	9.4–9.6	4.8–5.2
$NH4^+$-N	mg/L	20–24	0.12–0.27

[1] COD: Chemical oxygen demand; SS: Suspended solids; T-N: Total nitrogen; T-P: Total phosphorus. [2] IN: Influent; SE: Secondary effluent.

4.2. Samples Disinfected Using Chlorine or Ultraviolet Light

Chlorine-treated and ultraviolet-treated samples were collected from laboratory-scale batch disinfection experiments using secondary effluent samples (11/13, 11/20, and 11/27). All batch disinfection experiments employing chlorine or ultraviolet light were performed at room temperature. A free-chlorine stock solution was prepared in Milli-Q water with sodium hypochlorite (Wako, Japan) on the day of use. This stock solution was added to the secondary effluent samples (1000 mL each) at an initial free-chlorine concentration of 2 mg/L for 20 min, after which free-chlorine was neutralized immediately by adding sodium thiosulfate solution (Wako, Osaka, Japan). The residual free-chlorine concentrations were measured every 5 min using the *N,N*-diethyl-p-phenylenediamine method (Hach, Tokyo, Japan) to calculate concentration-time (CT) values. Free-chlorine CT values were the sum of the residual free-chlorine concentration (C) multiplied by the contact time (T) every 5 min for 20 min. The free-chlorine CT values of the chlorine-treated 11/13, 11/20, and 11/27 samples were 4.8, 2.9, and 2.3 mg·min/L, respectively.

A low-pressure ultraviolet lamp (ULO-6DQ; 254 nm; 6 W; Ushio, Tokyo, Japan) was used for laboratory-scale batch ultraviolet disinfection experiments. The ultraviolet lamp was stabilized before conducting experiments by turning it on for at least 40 min before use. The sample (500 mL) was added to sterilized glassware (Ushio) and exposed to ultraviolet light whilst stirring. Ultraviolet fluence was determined using an iodide–iodate actinometer [39,40]. Ultraviolet fluence values of ultraviolet-treated 11/13, 11/20, and 11/27 samples were 22, 30, and 21 mJ/cm^2, respectively.

4.3. IC–NGS Analysis of Infectious FRNAPH Strains

For the NGS analysis, 10 mL of influent samples and 100 mL of secondary effluent, chlorine-treated, and ultraviolet-treated samples were mixed with an equal volume of tryptone-glucose broth (10 g/L tryptone, 1.0 g/L glucose, 8.0 g/L NaCl, 0.3 g/L $CaCl_2$, 0.15 g/L $MgSO_4$, 20 mg/L kanamycin, and 100 mg/L nalidixic acid). The broth also contained *Salmonella enterica* serovar Typhimurium WG49, which was harvested during the exponential growth period and incubated at 37 °C overnight in order to propagate infectious FRNAPH strains. The propagated sample mixtures (15 mL) were centrifuged (2000 ×*g*, 10 min) and the supernatant was passed through a membrane filter (pore size 0.45 µm, hydrophilic cellulose acetate; Dismic-25cs, Advantec, Dublin, CA, USA) to remove bacteria, including the host strain. The filtrate (12 mL) was purified using a centrifugal filtration device (Amicon Ultra-15; Merck, Billerica, MA, USA) to increase the titres in the FRNAPH strains and remove soluble and low molecular weight components from the filtrate.

After purification, the samples (1 mL) were treated with RNase ONE Ribonuclease (Promega, Madison, WI, USA) (1 unit/50 µL of sample), and the mixture was incubated at 37 °C for 60 min to eliminate free RNA. Following RNase treatment, RNA was extracted using a QIAamp Viral RNA Mini QIAcube Kit (Qiagen, Hilden, Germany) and QIAcube (Qiagen), according to the manufacturer's protocol, followed by removal of DNA with Baseline-ZERO DNase (Arbrown, Chuo-ku, Japan). Bacterial ribosomal RNA was removed from the DNase-treated samples using a Ribo-Zero Bacteria Kit (Illumina, San Diego, CA, USA) according to the manufacturer's protocol. Libraries were then prepared using the TruSeq Stranded mRNA Library Prep Kit (Illumina), according to the manufacturer's protocol, without a purifying mRNA process. The TruSeq Stranded mRNA Library Prep Kit purifies poly(A)-containing mRNAs; however, the mRNAs of FRNAPHs do not contain poly(A) and are therefore excluded from this process. The libraries were subjected to agarose gel electrophoresis using E-Gel EX Agarose Gel (1%; Invitrogen, Carlsbad, CA, USA) with an E-Gel iBase Power System (Invitrogen). The cDNAs (300–600 bp) were then purified using a MonoFas DNA Purification Kit (GL Sciences, Torrance, CA, USA). The qualities and concentrations of purified cDNAs were assessed using an Agilent 2100 Bioanalyzer (Agilent Technologies, Santa Clara, CA, USA) and a Qubit Fluorometer (Invitrogen), respectively. The samples were pooled, and sequencing was performed using a MiSeq paired-end sequencing reaction with the v3 reagent kit (Illumina).

Before assembly of the metagenomic dataset, the quality of the MiSeq paired-end sequences was evaluated using FastQC then quality-trimmed and assembled de novo using Trimmomatic and Trinity, respectively, as implemented in the Galaxy platform (https://galaxy.dna.affrc.go.jp). Contigs >200 bp obtained from the de novo assembly were used as queries to perform a BLASTn version 2.7.1 + search with the NCBI nucleotide collection (nt) to identify significant alignments and the following parameters: A cut-off (e-value) of 10^{-3} and a maximum of one hit per read. The number of hits for FRNAPH strains was defined in order to count the number of FRNAPH strains identified as best hits according to the BLASTn analyses. The MEGAN program (version 6.12.0) was used to assign BLASTn hits for the taxonomy analysis.

4.4. *IC–RT-PCR–MPN Analysis of Infectious FRNAPH Genotypes*

IC–RT-PCR–MPN was performed to quantify the infectious FRNAPH genotypes as previously described [6,34,35]. Infectious FRNAPH genotypes in the samples were primarily propagated overnight at 37 °C by mixing with an equal volume of tryptone-glucose broth containing *S. enterica* WG49 (described above). Genotyping based on RT-PCR was subsequently applied, followed by quantification using the MPN method. The secondary effluent, chlorine-treated, and ultraviolet-treated samples were measured using sample volumes of 100, 10, 1, and 0.1 mL (n = 3 each). The detection limit of the secondary effluent, chlorine-treated, and ultraviolet-treated samples was 0.48 $\log_{10}$ MPN/L.

Supplementary Materials: The following are available online at http://www.mdpi.com/2076-0817/8/4/217/s1. Figure S1: Numbers of hits for FRNAPH strains representing each genotype in the 12 samples combined with a heat map showing the relative abundance of all FRNAPH strains according to the number of hits in the BLASTn analyses of influent (IN), secondary effluent (SE), chlorine-treated (Cl), and ultraviolet-treated (UV) samples. Blank cells indicate an absence of hits. Green and red cells indicate the lowest and highest values, respectively. Numbers in the heat-map cells indicate the number of hits for samples collected on 11/13, 11/20, and 11/27. Figure S2: Proportions of FRNAPH strains in each genotype combined with a heat map showing the relative abundance of each genotype according to the number of hits in the BLASTn analyses of influent (IN), secondary effluent (SE), chlorine-treated (Cl), and ultraviolet-treated (UV) samples. Blank cells indicate an absence of hits. White and blue (GI), sky blue (GI-JS), red (GII), green (GIII), and purple (GIV) cells indicate the lowest and highest values, respectively. Numbers in the heat-map cells indicate the proportions of FRNAPH strains in each genotype for samples collected on 11/13, 11/20, and 11/27.

Author Contributions: Conceptualization, S.L.; Methodology, S.L.; Validation, S.L., M.S., and H.S.; Investigation, S.L.; Data curation, S.L.; Writing—original draft preparation, S.L.; Writing—review and editing, M.S. and H.S.; Funding acquisition, S.L.

Funding: This research was funded by a KAKENHI grant (18K13863) from the Japan Society for the Promotion of Science (JSPS), Japan.

Conflicts of Interest: The authors declare no conflict of interest.

References

1. IAWPRC Study Group on Health Related Water Microbiology. Bacteriophages as model viruses in water quality control. *Water Res.* **1991**, *25*, 529–545. [CrossRef]
2. Grabow, W.O.K. Bacteriophages: Update on application as models for viruses in water. *Water SA* **2001**, *27*, 251–268. [CrossRef]
3. Bartsch, S.M.; Lopman, B.A.; Ozawa, S.; Hall, A.J.; Lee, B.Y. Global economic burden of norovirus gastroenteritis. *PLoS ONE.* **2016**, *11*, e0151219. [CrossRef] [PubMed]
4. Havelaar, A.; Van Olphen, M.; Drost, Y. F-specific RNA bacteriophages are adequate model organisms for enteric viruses in fresh water. *Appl. Environ. Microbiol.* **1993**, *59*, 2956–2962. [PubMed]
5. Hartard, C.; Leclerc, M.; Rivet, R.; Maul, A.; Loutreul, J.; Banas, S.; Boudaud, N.; Gantzer, C. F-specific RNA bacteriophages, especially members of subgroup II, should be reconsidered as good indicators of viral pollution of oysters. *Appl. Environ. Microbiol.* **2018**, *84*. [CrossRef] [PubMed]
6. Lee, S.; Suwa, M.; Shigemura, H. Occurrence and reduction of F-specific RNA bacteriophage genotypes as indicators of human norovirus at a wastewater treatment plant. *J. Water Health* **2019**, *17*, 50–62. [CrossRef] [PubMed]

7. Haramoto, E.; Otagiri, M.; Morita, H.; Kitajima, M. Genogroup distribution of F-specific coliphages in wastewater and river water in the Kofu basin in Japan. *Lett. Appl. Microbiol.* **2012**, *54*, 367–373. [CrossRef] [PubMed]
8. Haramoto, E.; Fujino, S.; Otagiri, M. Distinct behaviors of infectious F-specific RNA coliphage genogroups at a wastewater treatment plant. *Sci. Total Environ.* **2015**, *520*, 32–38. [CrossRef] [PubMed]
9. Hata, A.; Kitajima, M.; Katayama, H. Occurrence and reduction of human viruses, F-specific RNA coliphage genogroups and microbial indicators at a full-scale wastewater treatment plant in Japan. *J. Appl. Microbiol.* **2013**, *114*, 545–554. [CrossRef] [PubMed]
10. Muniesa, M.; Payan, A.; Moce-Llivina, L.; Blanch, A.R.; Jofre, J. Differential persistence of F-specific RNA phage subgroups hinders their use as single tracers for faecal source tracking in surface water. *Water Res.* **2009**, *43*, 1559–1564. [CrossRef] [PubMed]
11. Schaper, M.; Durán, A.E.; Jofre, J. Comparative resistance of phage isolates of four genotypes of F-specific RNA bacteriophages to various inactivation processes. *Appl. Environ. Microbiol.* **2002**, *68*, 3702–3707. [CrossRef] [PubMed]
12. Blatchley, E.R., III; Shen, C.; Scheible, O.K.; Robinson, J.P.; Ragheb, K.; Bergstrom, D.E.; Rokjer, D. Validation of large-scale, monochromatic UV disinfection systems for drinking water using dyed microspheres. *Water Res.* **2008**, *42*, 677–688. [CrossRef] [PubMed]
13. Hijnen, W.A.M.; Beerendonk, E.F.; Medema, G.J. Inactivation credit of UV radiation for viruses, bacteria and protozoan (oo)cysts in water: A review. *Water Res.* **2006**, *40*, 3–22. [CrossRef] [PubMed]
14. Friedman, S.D.; Cooper, E.M.; Casanova, L.; Sobsey, M.D.; Genthner, F.J. A reverse transcription-PCR assay to distinguish the four genogroups of male-specific (F+) RNA coliphages. *J. Virol. Methods* **2009**, *159*, 47–52. [CrossRef] [PubMed]
15. Friedman, S.D.; Cooper, E.M.; Calci, K.R.; Genthner, F.J. Design and assessment of a real time reverse transcription-PCR method to genotype single-stranded RNA male-specific coliphages (Family *Leviviridae*). *J. Virol. Methods* **2011**, *173*, 196–202. [CrossRef] [PubMed]
16. Friedman, S.D.; Snellgrove, W.C.; Genthner, F.J.; Division, G.E.; Breeze, G. Genomic sequences of two novel *Levivirus* single-stranded RNA coliphages (Family *Leviviridae*): Evidence for recombination in environmental strains. *Viruses* **2012**, 1548–1568. [CrossRef] [PubMed]
17. Furuse, K.; Sakurai, T.; Hirashima, A.; Katsuki, M.; Ando, A.; Watanabe, I. Distribution of ribonucleic acid coliphages in South and East Asia. *Appl. Environ. Microbiol.* **1978**, *35*, 995–1002. [PubMed]
18. Miyake, T.; Shiba, T.; Sakurai, T.; Watanabe, I. Isolation and properties of two new RNA phages SP and FI. *Jpn. J. Microbiol.* **1969**, *13*, 375–382. [CrossRef] [PubMed]
19. Hartard, C.; Rivet, R.; Banas, S.; Gantzer, C. Occurrence of and sequence variation among F-specific RNA bacteriophage subgroups in feces and wastewater of urban and animal origins. *Appl. Environ. Microbiol.* **2015**, *81*, 6505–6515. [CrossRef] [PubMed]
20. Langlet, J.; Gaboriaud, F.; Duval, J.F.L.; Gantzer, C. Aggregation and surface properties of F-specific RNA phages: Implication for membrane filtration processes. *Water Res.* **2008**, *42*, 2769–2777. [CrossRef] [PubMed]
21. Boudaud, N.; Machinal, C.; David, F.; Fréval-Le Bourdonnec, A.; Jossent, J.; Bakanga, F.; Arnal, C.; Jaffrezic, M.P.; Oberti, S.; Gantzer, C. Removal of MS2, Qβ and GA bacteriophages during drinking water treatment at pilot scale. *Water Res.* **2012**, *46*, 2651–2664. [CrossRef] [PubMed]
22. Dika, C.; Ly-Chatain, M.H.; Francius, G.; Duval, J.F.L.; Gantzer, C. Non-DLVO adhesion of F-specific RNA bacteriophages to abiotic surfaces: Importance of surface roughness, hydrophobic and electrostatic interactions. *Colloids Surf. A Physicochem. Eng. Asp.* **2013**, *435*, 178–187. [CrossRef]
23. Schaper, M.; Jofre, J.; Uys, M.; Grabow, W.O.K. Distribution of genotypes of F-specific RNA bacteriophages in human and non-human sources of faecal pollution in South Africa and Spain. *J. Appl. Microbiol.* **2002**, *92*, 657–667. [CrossRef] [PubMed]
24. Cole, D.; Long, S.C.; Sobsey, M.D. Evaluation of F+ RNA and DNA coliphages as source-specific indicators of fecal contamination in surface waters. *Appl. Environ. Microbiol.* **2003**, *73*, 22–23. [CrossRef] [PubMed]
25. Ogorzaly, L.; Tissier, A.; Bertrand, I.; Maul, A.; Gantzer, C. Relationship between F-specific RNA phage genogroups, faecal pollution indicators and human adenoviruses in river water. *Water Res.* **2009**, *43*, 1257–1264. [CrossRef] [PubMed]

26. Wolf, S.; Hewitt, J.; Rivera-Aban, M.; Greening, G.E. Detection and characterization of F+ RNA bacteriophages in water and shellfish: Application of a multiplex real-time reverse transcription PCR. *J. Virol. Methods* **2008**, *149*, 123–128. [CrossRef] [PubMed]
27. Haramoto, E.; Kitajima, M.; Katayama, H.; Asami, M.; Akiba, M.; Kunikane, S. Application of real-time PCR assays to genotyping of F-specific phages in river water and sediments in Japan. *Water Res.* **2009**, *43*, 3759–3764. [CrossRef] [PubMed]
28. Gourmelon, M.; Caprais, M.P.; Mieszkin, S.; Marti, R.; Wéry, N.; Jardé, E.; Derrien, M.; Jadas-Hécart, A.; Communal, P.Y.; Jaffrezic, A.; et al. Development of microbial and chemical MST tools to identify the origin of the faecal pollution in bathing and shellfish harvesting waters in France. *Water Res.* **2010**, *44*, 4812–4824. [CrossRef] [PubMed]
29. Mieszkin, S.; Caprais, M.P.; Le Mennec, C.; Le Goff, M.; Edge, T.A.; Gourmelon, M. Identification of the origin of faecal contamination in estuarine oysters using Bacteroidales and F-specific RNA bacteriophage markers. *J. Appl. Microbiol.* **2013**, *115*, 897–907. [CrossRef] [PubMed]
30. Cantalupo, P.G.; Calgua, B.; Zhao, G.; Hundesa, A.; Wier, A.D.; Katz, J.P.; Grabe, M.; Hendrix, R.W.; Girones, R.; Wang, D.; et al. Raw sewage harbors diverse viral populations. *MBio* **2011**, *2*. [CrossRef] [PubMed]
31. Tamaki, H.; Zhang, R.; Angly, F.E.; Nakamura, S.; Hong, P.Y.; Yasunaga, T.; Kamagata, Y.; Liu, W.T. Metagenomic analysis of DNA viruses in a wastewater treatment plant in tropical climate. *Environ. Microbiol.* **2012**, *14*, 441–452. [CrossRef] [PubMed]
32. Bibby, K.; Peccia, J. Identification of viral pathogen diversity in sewage sludge by metagenome analysis. *Environ. Sci. Technol.* **2013**, *47*, 1945–1951. [CrossRef] [PubMed]
33. Aw, T.G.; Howe, A.; Rose, J.B. Metagenomic approaches for direct and cell culture evaluation of the virological quality of wastewater. *J. Virol. Methods* **2014**, *210*, 15–21. [CrossRef] [PubMed]
34. Hata, A.; Hanamoto, S.; Shirasaka, Y.; Yamashita, N.; Tanaka, H. Quantitative distribution of infectious F-specific RNA phage genotypes in surface waters. *Appl. Environ. Microbiol.* **2016**, *82*, 4244–4252. [CrossRef] [PubMed]
35. Lee, S.; Tasaki, S.; Hata, A.; Yamashita, N.; Tanaka, H. Evaluation of virus reduction at a large-scale wastewater reclamation plant by detection of indigenous F-specific RNA bacteriophage genotypes. *Environ. Technol.* **2019**, *40*, 2527–2537. [CrossRef] [PubMed]
36. Bömer, M.; Rathnayake, A.I.; Visendi, P.; Sewe, S.O.; Paolo, J.; Sicat, A.; Silva, G.; Kumar, P.L.; Seal, S.E. Tissue culture and next-generation sequencing: A combined approach for detecting yam (*Dioscorea* spp.) viruses. *Physiol. Mol. Plant Pathol.* **2019**, *105*, 54–66. [CrossRef] [PubMed]
37. Haft, D.H.; Tovchigrechko, A. High-speed microbial community profiling. *Nat. Methods* **2012**, *9*, 793–794. [CrossRef] [PubMed]
38. Tan, B.F.; Ng, C.; Nshimyimana, J.P.; Loh, L.L.; Gin, K.Y.H.; Thompson, J.R. Next-generation sequencing (NGS) for assessment of microbial water quality: Current progress, challenges, and future opportunities. *Front. Microbiol.* **2015**, *6*, 1027. [CrossRef] [PubMed]
39. Rahn, R.O. Potassium iodide as a chemical actinometer for 254 nm radiation: Use of iodate as an electron scavenger. *Photochem. Photobiol.* **1997**, *66*, 450–455. [CrossRef]
40. Rahn, R.O.; Stefan, M.I.; Bolton, J.R.; Goren, E.; Shaw, P.S.; Lykke, K.R. Quantum yield of the iodide-iodate chemical actinometer: Dependence on wavelength and concentrations. *Photochem. Photobiol.* **2003**, *78*, 146–152. [CrossRef]

MDPI
St. Alban-Anlage 66
4052 Basel
Switzerland
Tel. +41 61 683 77 34
Fax +41 61 302 89 18
www.mdpi.com

Pathogens Editorial Office
E-mail: pathogens@mdpi.com
www.mdpi.com/journal/pathogens

www.ingramcontent.com/pod-product-compliance
Lightning Source LLC
LaVergne TN
LVHW072001080526
838202LV00064B/6811
* 9 7 8 3 0 3 9 2 8 5 6 8 6 *